ADVANCED MOLECULAR DYNAMICS AND CHEMICAL KINETICS

ADVANCED MOLECULAR DYNAMICS AND CHEMICAL KINETICS

GERT D. BILLING
KURT V. MIKKELSEN
Department of Chemistry
H. C. Ørsted Institute
University of Copenhagen

A Wiley-Interscience Publication
JOHN WILEY & SONS, INC.
New York • Chichester • Weinheim • Brisbane • Singapore • Toronto

Phys

Copyright © 1997 by John Wiley & Sons, Inc.

All rights reserved. Published simultaneously in Canada.

Library of Congress Cataloging in Publication Data:
Billing, Gert D.
    Advanced molecular dynamics and chemical kinetics / Gert D.
Billing and Kurt V. Mikkelsen.
        p.   cm.
    "A Wiley-Interscience publication."
    Includes index.
    ISBN 0-471-12740-X (cloth : alk. paper)
    1. Chemical kinetics.   2. Molecular dynamics.   I. Mikkelsen, Kurt
V. II. Title.
QD461.B5725   1997
541.3′9—dc20   96-43762
                                                                  CIP

Printed in the United States of America

10 9 8 7 6 5 4 3 2 1

To Maryanne and Inge-Lise

CONTENTS

PREFACE

This monograph is intended for the young researcher at the Ph.D. or postdoctoral level who wishes to enter the field of molecular reaction dynamics or perhaps simply wishes to learn more about it. Our hope is that those already in the field will be able to use it as a reference. The material in this book has evolved from teaching notes used and revised in several courses for graduate students and postdoctoral candidates held over the last few years at the University of Copenhagen and Aarhus University. Furthermore, we have used the material for a two-week international summer school in molecular dynamics and chemical kinetics held at Fuglsøcentret, Knebel, Denmark, in August 1995. It is our intention to continue to hold this summer school every other year, using material from this book and our previous book (*Introduction to Molecular Dynamics and Chemical Kinetics*), along with updated and new material. Both of us have used parts of the material for short Ph.D. courses at other universities: University of Bari and Uppsala University.

Our intention has been to present a selection of methods that, during the last few decades, have turned out to be interesting. However, what is included here is but a selection, colored by personal experience and preferences. Nevertheless, the book contains a great many subjects—perhaps too many to be covered in a single (semester) course. Some topics are presented, if only briefly, because we consider it to be important to have at the very least a superficial knowledge of some of these techniques and to know that such methods exist. The ever-continuing development in the field of molecular reaction dynamics will tell whether our selection has been wise.

We have mainly focused on time-dependent methods, the reason being that these methods are somewhat easier to implement and to understand, as well as to use, than time-independent methods, though they are not in general faster from a numerical point of view. Furthermore, time-dependent methods make the connection to approximate, i.e., semiclassical, methods more obvious.

The nature of the subjects covered in the text make simple exercises impossible. Instead, the methods must be implemented numerically. We have, however, included exercises whenever possible. (The answers to these are given in an appendix.)

After the introduction (Chapter 1), the monograph starts out with a chapter on second quantization methods, Chapter 2. This formulation is widely used in many areas of quantum dynamics. We include some examples on energy transfer to polyatomic molecules and solids, and quantum mechanical radiation fields. Chapter 3 deals with effective Hamiltonians and the time-dependent self-consistent field (SCF) method. Semiclassical methods are introduced in Chapter 4. Wavepacket propagation and grid methods have become popular tools in molecular reaction dynamics; they are discussed in Chapter 5. Potential energy surfaces and the Born-Oppenheimer separation are discussed briefly in Chapter 6. The reaction path and variational transition state methods are reviewed in Chapters 7 and 8. In Chapter 9 the flux–flux correlation method for chemical reactions is derived. Phase space and stochastic theories are introduced in Chapter 10. Photodissociation is discussed within a time-dependent framework in Chapter 11. The concept of density operators is introduced in Chapter 12, and subsequently, in Chapter 13, we use density operators along with equation of motion approaches for describing how a molecular system evolves while coupled to a heat bath. In Chapter 14 we focus on how a condensed phase changes the molecular energetics of an immersed molecular system. Chapter 15 presents response methods for time-dependent molecular properties of solvated molecules, while Chapter 16 contains a model for calculating magnetic properties of solvated molecules. Electron transfer reactions in solution are the topic of Chapter 17, and in Chapter 18 we discuss methods for calculating electron transfer coupling elements. In Chapter 19, we consider a method for investigating the influence of quantum mechanical radiation on charge transfer reactions. The final chapter, Chapter 20, concerns models for proton transfer reactions in solution.

In order to be able to benefit from this book, the reader must have prior knowledge. We find that a natural progression would be our first book, *Introduction to Molecular Dynamics and Chemical Kinetics*; the book by G. Schatz and M. A. Ratner, *Quantum Mechanics in Chemistry*; followed by the present book. These three books span the introductory level to research in the field. We expect that the reader, after finishing this book, and thereby obtaining the necessary concepts and ideas, would be able to plunge into the research pool of molecular dynamics and chemical kinetics.

We acknowledge the stimulating and worthy collaborations that we have had through the years with quite a large number of people. Furthermore, we thank our students for their suggestions and criticism. Their contributions and suggestions for expanding certain areas in the book are gratefully acknowledged. Maryanne Kmit's review of the manuscript is also gratefully acknowledged.

GERT D. BILLING
KURT V. MIKKELSEN

Copenhagen, 1996

1

INTRODUCTION

The purpose of the present monograph *Advanced Molecular Dynamics and Chemical Kinetics*, is to give an introduction to methods that are used today in this very active research field. The development of advanced beam and laser experiments has made a large body of experimental data available, providing information on chemical reactivity and energy transfer processes at the molecular level. In some cases the interpretation of these data is straightforward—but increasingly, because of the complexity of the experiments and the chemical processes involved, the support and interpretation offered by reliable dynamical calculations is vital. Furthermore, information on many dynamical processes at the "state resolved" level is needed for simulating bulk phenomena. This type of information is often not accessible experimentally. Hence dynamical calculations are even more important in such cases.

Traditionally calculations are carried out using an approach consisting of three parts:

(1) Born-Oppenheimer separation
(2) Construction of the potential energy surface
(3) Dynamical calculations

The approach is based upon the use of the Born-Oppenheimer (BO) separation of the nuclear and electronic motion (see Chapter 6). However, it should be mentioned that many processes involve such strong coupling between these two degrees of freedom that this standard procedure will be neither the most convenient nor the correct way of proceeding. Nevertheless if the BO separation is introduced, the next goal is the construction of suitable potential energy surfaces (one or more) for the nuclear dynamics. If more than one surface is involved, nonadiabatic coupling terms are also required. In principle this is possible using *ab initio* electronic structure calculations. Due to the complexity of the many-electron problem (the problem at hand when we are concerned with interesting chemical processes) and the fact that a large part of the nuclear configuration space has to be covered, such calculations are not yet performed routinely. Instead most potential energy surfaces are constructed semiempirically using *ab initio* information when available. The next step required is the construction of analytical potential-energy-surface expressions. In this phase it is necessary to include information from sources such as spectroscopy, experimentally or theoretically determined long-range interactions, measurements of transport properties and barrier heights. The construction of analytical potential surfaces is a discipline of its own and no standard methodology is available at present (for further reading see refs. [1, 2]). Once the potential surface is constructed, dynamical calculations can begin. By removing the center of mass motion and assuming that the total angular momentum and its projection on a space-fixed axis are conserved, the exact solution of the nuclear motion is reduced to a $3N - 5$ dimensional problem, where N is the number of particles. Unless $N \sim 3$ to 4, it is also necessary to introduce approximations at this stage. However, since the nuclear masses are much larger than the mass of the electrons, it is possible, without significant errors, to introduce an approximate description of all, or at least part, of the nuclear motion. One obvious choice is to use classical dynamics; however, quantum-classical (semiclassical) techniques are in many cases more appropriate. The reason for this is that the nuclear motion is subject to various quantum phenomena such as tunneling, zero-point vibrational energy, nonadiabatic electronic processes, interference, and resonances. The classical limit of quantum mechanics can conveniently be studied using the Feynman path integral formalism (Chapter 4).

Methods which mix quantum and classical mechanics offer enough flexibility to treat a significant portion of the processes in reaction dynamics (see Chapters 7 and 11). For dissociative and reaction coordinates it is convenient to introduce a discrete representation of the wave function, i.e., to use grid methodology (Chapter 5). Methods which do not require complete knowledge of the potential energy surface are the reaction path (Chapter 7), and the variational and quantum transition state methods (Chapters 8 and 9). One particularly interesting problem is how the environment influences reactivity, be it influence from a solvent, an external field, a "rigid" part of the reacting molecules or catalytic effects from metals or active sites on proteins. In order to treat such processes we need to obtain information on how the molecular properties change when the molecular system is interacting either with solvent molecules, surfaces, or electromagnetic fields. In order to describe these many-body problems one has to introduce new concepts and methods designed for handling these complex problems, and we introduce density operators and reduced density operators in order to focus on the system of interest and have a less detailed description of the (chemically) less interesting degrees of freedom (Chapters 12 and 13). Prior to developing models for chemical reactions in solution, it is necessary to establish theoretical and numerical models for investigating how the solvent affects the molecular properties of solvated molecules. Chapters 14 through 16 illustrate approaches for determining electric and magnetic molecular properties of solvated compounds. Charge transfer reactions are strongly influenced by the presence of the solvent and we address this class of reactions in Chapter 17. Nonadiabatic coupling elements are of importance in all chemical reactions; we outline approaches for calculating these in the case of electron transfer reactions (Chapter 18). The effects of quantum mechanical radiation fields depend strongly on the pulse shape and pulse delay of the applied field and in most situations it is necessary to combine various methods in order to investigate these effects for solvated molecular systems (Chapter 19). Proton transfer reactions in solution are another prototype of charge transfer reactions, and have recently become a quite active research area (Chapter 20).

The molecular processes dealt with in the present monograph encompass a large variety, ranging from simple energy transfer processes in the gas phase, to solvent effects on reactions in liquids, and from small to large molecular systems. Such variety and richness within the field of molecular dynamics requires the introduction of many different methods—from exact quantum mechanical descriptions of small systems to approximate statistical methods for large systems (see Table 1.1).

The introduction of the BO separation makes it possible to obtain information needed for dynamical simulations from sources other than electronic structure calculations, for example, from spectroscopy, beam data, semiempirical calculations, or models. This is fortunate in the sense that the dynamical methods are at present capable of treating systems consisting of a large number of atoms—up to several thousand if classical mechanics is assumed for the nuclear motion. However if the nuclear dynamics is solved using a full quantum description, only systems of about 3 to 4 atoms can be treated with the present computer facilities. From a computational point of view, it is important

TABLE 1.1. Overview of Methods and Descriptions in Molecular Dynamics[a]

Description	Dynamics	Potential
Quantum	Coupled eqs., coupled wavepackets.	Full
Semiclassical	Trajectories, Gaussian wavepackets, coupled equations	Full
Classical	Trajectories	Full
Reaction path	Wavepackets, trajectories	Restricted
Transition state	Counting of states, reactive flux	Restricted
Phase space/statistical	Counting of states	Restricted

[a]Some need only restricted information for the potential energy surface, others need the full potential.

that some of the methods are aimed at obtaining much less information than, say, state- and energy-resolved differential cross sections. These methods aim at obtaining rate constants as a function of temperature (transition state theory) or also at obtaining state-resolved rates (the reaction path method). It then follows that these methods require less information on the potential energy surface, that is information around the transition state and along the reaction path. If very little information is available on the interaction potential, one can estimate reaction cross sections using statistical or phase space methods.

2

SECOND QUANTIZATION

A description of many-particle systems requires the inclusion of interparticle potentials in the many-particle Schrödinger equation (SE). In principle, the N-body wave function in configuration space contains all possible information, but a direct solution is impossible. Second quantization (SQ) greatly simplifies the discussion of many identical interacting particles. In relativistic theory the concept of SQ is essential for describing the creation and destruction of particles. Second quantization reformulates the original SE and has the advantage of incorporating the statistics at each step, whether the particle is a fermion or a boson. This is an advantage compared to the more cumbersome approach using symmetric and antisymmetric products of single-particle wavefunctions. The methods of SQ also allow us to concentrate on the few matrix elements of interest, thus avoiding the need for dealing directly with the many-particle wavefunction and the coordinates of all the remaining particles.

The Hamiltonian for an N-particle system where one- and two-particle interaction is included, takes the form

$$H = \sum_{k=1}^{N} T(x_k) + \frac{1}{2} \sum_{k,l;k \neq l}^{N} V(x_k, x_l) \qquad (2.1)$$

where T is the kinetic energy, V the potential energy between the particles, and x_k the coordinates of the kth particle, i.e., it includes the spatial coordinates and any discrete variables, such as the spin for a system of fermions.

The potential energy term represents the interaction between every pair of particles, counted once. This accounts for the factor $\frac{1}{2}$ when the double sum runs over indices k and l (excluding $k = l$). The time-dependent Schrödinger equation (TDSE) is

$$i\hbar \frac{\partial}{\partial t} \Psi(x_1, \ldots, x_k, \ldots, x_N, t) = H\Psi(x_1, \ldots, x_k, \ldots, x_N, t) \qquad (2.2)$$

together with an appropriate set of boundary conditions for the wave function. We now expand the many-particle wavefunction in a complete set of time-independent single-particle wave functions $\Psi_{E_k}(x_k)$ that incorporate the boundary conditions. Here E_k represents the complete set of single-particle quantum numbers. We let this infinite set of single-particle quantum numbers be ordered as $(1, 2, 3 \ldots r, s, t \ldots \infty)$ and note that E_k runs over this set of eigenvalues.

The many-body wave function can then be expanded as

$$\Psi(x_1, \ldots, x_k, \ldots, x_N, t) = \sum_{E'_1, \ldots, E'_N} C(E'_1, \ldots, E'_N, t)\Psi_{E'_1}(x_1) \ldots \Psi_{E'_N}(x_N) \quad (2.3)$$

This expression is completely general and is simply the expansion of the many-particle wave function in a complete set of functions. Since the single-particle wave functions are time-independent, the time dependence appears in the coefficients $C(E'_1, \ldots, E'_N, t)$. Let us now insert this expansion in the TDSE and multiply by $\Psi_{E_1}(x_1)^* \ldots \Psi_{E_N}(x_N)^*$, i.e., a product of the complex conjugate wave functions corresponding to a fixed set of quantum numbers $E_1, \ldots E_N$. We then obtain:

$$i\hbar \Psi_{E_1}(x_1)^* \ldots \Psi_{E_N}(x_N)^* \frac{\partial}{\partial t} \sum_{E'_1, \ldots, E'_N} C(E'_1, \ldots, E'_N, t)\Psi_{E'_1}(x_1) \ldots \Psi_{E'_N}(x_N)$$

$$= \Psi_{E_1}(x_1)^* \ldots \Psi_{E_N}(x_N)^* H \sum_{E'_1, \ldots, E'_N} C(E'_1, \ldots, E'_N, t)\Psi_{E'_1}(x_1) \ldots \Psi_{E'_N}(x_N)$$

$$(2.4)$$

Integrating over appropriate coordinates $\langle \Psi_{E_i} | \Psi_{E_j} \rangle = \delta_{ij}$ yields on the left hand side $i\hbar \partial/\partial t\, C(E_1, \ldots, E_N, t)$. The kinetic energy operator is a single-particle operator involving the coordinates of the particles one at a time. It can change only one of the single-particle wave functions. The orthonormality of the single-particle wave functions ensures that all but the kth particle must have the original given quantum numbers. The wave function of the kth particle can still run over the infinite set of quantum numbers. We then obtain for the kinetic energy operator

$$\sum_{k=1}^{N} \sum_{E_\omega} C(E_1, \ldots, E_{k-1}, E_\omega, E_{k+1}, \ldots, E_N, t) \int dx_k \Psi_{E_k}(x_k)^* T(x_k) \Psi_{E_\omega}(x_k)$$

$$(2.5)$$

Unlike the kinetic energy operator, the potential energy operator in Eq. (2.1) involves the coordinates of two particles and can, at most, change the wave functions of two particles e.g., the kth and the lth, leaving all the other quantum numbers unchanged. The quantum number of the kth particle still runs over an infinite set of values (denoted E_ω). The same holds for the lth particle (denoted $E_{\omega'}$). We obtain

$$\frac{1}{2} \sum_{k \neq l=1}^{N} \sum_{E_\omega} \sum_{E_{\omega'}} C(E_1, \ldots, E_{k-1}, E_\omega, E_{k+1}, \ldots, E_{l-1}, E_{\omega'}, E_{l+1}, \ldots, E_N, t)$$

$$\int dx_k \, dx_l \, \Psi_{E_k}(x_k)^* \Psi_{E_l}(x_l)^* V(x_k, x_l) \Psi_{E_\omega}(x_k) \Psi_{E_{\omega'}'}(x_l) \qquad (2.6)$$

Each set of quantum numbers leads to a different equation, yielding an infinite set of coupled differential equations for the time-dependent coefficient of the many-particle wave function. We now wish to incorporate the proper statistics of the particles, i.e. that

$$\Psi(x_1, \ldots, x_j, \ldots, x_l, \ldots, x_N, t) = \pm \Psi(x_1, \ldots, x_l, \ldots, x_j, \ldots, x_N, t) \qquad (2.7)$$

i.e., the wave function must be either symmetric (bosons) or antisymmetric (fermions) under the exchange of coordinates of any two particles. A necessary and sufficient condition for this to hold is that the C's must be either symmetric or antisymmetric under interchange of the corresponding quantum numbers

$$C(E_1, \ldots, E_k, \ldots, E_l, \ldots, E_N, t) = \pm C(E_1, \ldots, E_l, \ldots, E_k, \ldots, E_N, t) \qquad (2.8)$$

where $+$ holds for bosons and $-$ for fermions.

We have illustrated the possibility of specifying states and statistics in terms of functions/coefficients that govern quantum numbers instead of particle coordinates and wave functions [these being integrated out in Eq. (2.6)]. Let us now consider a somewhat different quantum mechanical basis that describes the number of particles (or quanta) occupying each state in a complete set of single-particle states. We have time-independent abstract state vectors

$$|n_1 n_2 \ldots n_\infty\rangle = |n_1\rangle |n_2\rangle \ldots |n_\infty\rangle \qquad (2.9)$$

where n_1 denotes the number of particles in eigenstate 1. We require that this basis be complete and orthonormal. This requires that these states satisfy

$$\langle n'_1 n'_2 \ldots n'_\infty | n_1 n_2 \ldots n_\infty \rangle = \delta_{n'_1 n_1} \delta_{n_2 n'_2} \ldots \delta_{n'_\infty n_\infty} \qquad (2.10)$$

and

$$\sum_{n_1 n_2 \ldots n_\infty} |n_1 n_2 \ldots n_\infty \rangle \langle n_1 n_2 \ldots n_\infty | = 1 \qquad (2.11)$$

Let us furthermore introduce the zero particle state or vacuum state $| \ vac \ \rangle$. In order to be more concrete we introduce the time-independent operators $b_k, b_k^\dagger$ that satisfy the following commutation relations:

$$[b_{k'}, b_k^\dagger] = b_{k'} b_k^\dagger - b_k^\dagger b_{k'} = \delta_{k'k} \qquad (2.12)$$

and

$$[b_{k'}, b_k] = [b_{k'}^\dagger, b_k^\dagger] = 0 \qquad (2.13)$$

These are just the commutation rules for the creation and destruction operators for boson particles. Other relations include

$$b_k^\dagger b_k |n_k\rangle = n_k |n_k\rangle \qquad (2.14)$$

$$b_k |n_k\rangle = \sqrt{n_k} |n_k - 1\rangle \qquad (2.15)$$

$$b_k^\dagger |n_k\rangle = \sqrt{n_k + 1} |n_k + 1\rangle \qquad (2.16)$$

where $b_k^\dagger b_k$ is the number operator, i.e., when operating on the state k, it gives n_k the number of particles in that specific state. The destruction operator b_k decreases the occupation number by 1 and multiplies it with $\sqrt{n_k}$, whereas the creation operator $b_k^\dagger$ increases the occupation number by 1 and multiplies the state by $\sqrt{n_k + 1}$.

As a simple illustration consider the eigenvalues of the number operator. First we note that the operator is Hermitian, $(b_k^\dagger b_k)^\dagger = b_k^\dagger b_k$ and therefore the eigenvalues are real. We can also deduce that these eigenvalues are greater than or equal to zero. Thus we have

$$n = \langle n|b^\dagger b|n\rangle = \sum_m \langle n|b^\dagger + |m\rangle \langle m|b|n\rangle = \sum_m |\langle m|b|n\rangle|^2 \qquad (2.17)$$

which shows that n is either zero or positive.

Consider also the commutation relation

$$[b^\dagger b, b] = b^\dagger bb - bb^\dagger b = [b^\dagger, b]b = -b \tag{2.18}$$

Using this we can see how b acts on an eigenstate with eigenvalue n

$$b^\dagger b(b|n\rangle) = bb^\dagger b|n\rangle + [b^\dagger b, b]|n\rangle = nb|n\rangle - b|n\rangle$$
$$= (n-1)b|n\rangle \tag{2.19}$$

Repeated use of b on any eigenstate must eventually give zero, since we otherwise could create a state with negative eigenvalue of the number operator, which contradicts Eq. (2.17). Therefore zero is one possible eigenvalue of the number operator. The adjoint commutation relation shows

$$[b^\dagger b, b^\dagger] = b^\dagger bb^\dagger - b^\dagger b^\dagger b = b^\dagger[b, b^\dagger] = b^\dagger \tag{2.20}$$

Following the above procedure we obtain

$$b^\dagger b(b^\dagger|n\rangle) = b^\dagger b^\dagger b|n\rangle + [b^\dagger b, b^\dagger]|n\rangle = nb^\dagger|n\rangle + b^\dagger|n\rangle$$
$$= (n+1)b^\dagger|n\rangle \tag{2.21}$$

i.e., the operator $b^\dagger$ is a creation operator and increases the eigenvalues of the number operator by one. The number operators for different modes commute, which means that the eigenstates of the total system can simultaneously be eigenstates of the set $\{b_k^\dagger b_k\}$. Our occupation number basis states are simply the direct product of eigenstates of the number operator for each state.

When considering the fermion operator, we must take into account the antisymmetry under interchange of quantum numbers. This implies that the quantum number, E_l, must be different from the quantum number, E_k, otherwise the coefficient is zero. This result shows that the occupation number, n_k, must be either zero or one, reflecting the Pauli principle.

Any coefficients that have the same states occupied are equal except for a minus sign. Therefore, we will define our coefficients in such a way that

$$C(n_1, \ldots, n_k, \ldots, n_l, \ldots, n_N, t) = C(E_1 \langle \ldots \langle E_l \langle \ldots \langle E_k \langle \ldots E_N, t) \tag{2.22}$$

where we first arrange all the quantum numbers in the coefficient in an increasing sequence. We now introduce the abstract occupation number space and the anticommutation rules for fermions

$$\{a_r, a_s^\dagger\} = \delta_{rs} \tag{2.23}$$
$$\{a_r, a_s\} = \{a_r^\dagger, a_s^\dagger\} = 0 \tag{2.24}$$

where the anticommutator is evaluated as

$$\{A, B\} = [A, B]_+ = AB + BA \tag{2.25}$$

It can easily be seen from this set of commutators, that we get the correct statistics:

(1) $(a_s)^2 = (a_s^\dagger)^2$, therefore $a_s^\dagger a_s^\dagger \mid \text{vac}\rangle = 0$, which prevents two particles from occupying the same state.

(2) Operators for the same mode $a^\dagger a = 1 - aa^\dagger$ and $(a^\dagger a)^2 = 1 - 2aa^\dagger + aa^\dagger aa^\dagger = 1 - 2aa^\dagger + a(1 - aa^\dagger)a^\dagger = 1 - 2aa^\dagger + aa^\dagger - aaa^\dagger a^\dagger = 1 - aa^\dagger = a^\dagger a$ or $a^\dagger a(1 - a^\dagger a) = 0$. This implies that the number operator for a given mode has eigenvalues zero and one as required.

(3) The anticommutation rules themselves along with an overall choice of phase yield the following relations for the raising and lowering operators

$$a^\dagger \mid \text{vac}\rangle = \mid 1\rangle \qquad a \mid 1\rangle = \mid \text{vac}\rangle \tag{2.26}$$
$$a \mid \text{vac}\rangle = 0 \qquad a^\dagger \mid 1\rangle = 0 \tag{2.27}$$

The anticommutation rules slightly complicate the direct-product state $\mid n_1 n_2 \ldots n_\infty\rangle$ because it becomes essential to keep track of signs. With the definition

$$\mid n_1 n_2 \ldots n_\infty\rangle = (a_1^\dagger)^{n_1} (a_2^\dagger)^{n_2} \ldots (a_\infty^\dagger)^{n_\infty} \mid \text{vac}\rangle \tag{2.28}$$

we can evaluate directly the effect of the destruction operator, a_s on this state

$$a_s \mid n_1 n_2 \ldots n_\infty\rangle = a_s (a_1^\dagger)^{n_1} \ldots (a_\infty^\dagger)^{n_\infty} \mid \text{vac}\rangle$$
$$= (-1)^{n_1 + n_2 + \ldots + n_{s-1}} (a_1^\dagger)^{n_1} \ldots (a_s a_s^\dagger) \ldots (a_\infty^\dagger)^{n_\infty} \mid \text{vac}\rangle \tag{2.29}$$

if $n_s = 0$ the operator can be moved all the way to the vacuum state $\mid \text{vac}\rangle$ yielding zero. If there is one particle in the state n_s, it is convenient first to use the anticommutation relation

$$a_s a_s^\dagger = 1 - a_s^\dagger a_s \tag{2.30}$$

and then take the a_s of the second term to operate on $\mid \text{vac}\rangle$ giving zero. We have

$$a_s \mid n_1 \ldots n_s \ldots n_\infty\rangle = n_s (-1)^{n_1 + \ldots + n_{s-1}} \mid n_1 \ldots n_s - 1 \ldots n_\infty\rangle \tag{2.31}$$

for $n_s = 1$, otherwise 0

$$a_s^\dagger |n_1 \ldots n_s \ldots n_\infty\rangle = (1 - n_s)(-1)^{n_1 + \ldots + n_{s-1}} |n_1 \ldots n_s + 1 \ldots n_\infty\rangle \qquad (2.32)$$

for $n_s = 0$, otherwise 0.

$$a_s^\dagger a_s |n_1 \ldots n_s \ldots n_\infty\rangle = n_s |n_1 \ldots n_s \ldots n_\infty\rangle \qquad (2.33)$$

The SE for a many-electron system in the SQ formulation can be written as [14]

$$i\hbar \frac{\partial}{\partial t} |\Psi(t)\rangle = H|\Psi(t)\rangle \qquad (2.34)$$

with

$$H = \sum_{rs} \langle r|T|s\rangle a_r^\dagger a_s + \frac{1}{2} \sum_{k \neq l = 1} \langle rs|V|tu\rangle a_r^\dagger a_s^\dagger a_u a_t \qquad (2.35)$$

This is equivalent to the SE in the first quantization—the wave function approach—but the phases arising from the antisymmetry of the expansion coefficients have been incorporated in the Hamiltonian and the direct-product state vectors. For SQ used in conjunction with fermions, the ordering is important, since a change in the order of the operators in the latter term introduces sign changes. However, in the case of bosons the ordering is irrelevant.

In order to rationalize Eq. (2.35) we consider a one-particle operator $H(x, \hat{p}_x)$ where

$$H_{ij} = \langle \psi_i | H | \psi_j \rangle = \int dx \, \psi_i^*(x) H(x, \hat{p}_x) \psi_j(x) \qquad (2.36)$$

which in the second quantization formulation is expressed as

$$H = \sum_{ij} H_{ij} a_i^\dagger a_j \qquad (2.37)$$

since

$$\langle 0 \ldots 1(i) \ldots |H| 0 \ldots 1(j) \ldots \rangle = \sum_{kl} H_{kl} \langle \ldots |a_k^\dagger a_l| \ldots \rangle$$

$$= \sum_{kl} H_{kl} \delta_{lj} \delta_{ik} = H_{ij} \qquad (2.38)$$

where the index in parentheses indicates the state number. Similarly for the N-particle case where

$$H = \sum_{i=1}^{N} H(x_i \hat{p}_i) \tag{2.39}$$

the occupation number formalism again yields Eq. (2.37). Examples of one-particle operators are the kinetic energy operator $-\hbar^2/2m(\partial^2/\partial x^2)$ or the Bloch operator

$$-\frac{\hbar^2}{2m} \frac{\partial^2}{\partial x^2} + V(x) \tag{2.40}$$

where $V(x)$ is a periodic potential.

A two-particle operator is, e.g., the Coulomb potential $e^2/|x_i - x_j|$. For a Hamiltonian consisting of two-particle operators

$$H = \frac{1}{2} \sum_{i \neq j}^{N} H(x_i \hat{p}_i x_j \hat{p}_j) \tag{2.41}$$

we get

$$H = \frac{1}{2} \sum_{klmn} H_{klmn} a_l^{\dagger} a_k^{\dagger} a_m a_n \tag{2.42}$$

where

$$H_{klmn} = \int dx \int dx' \ \psi_k^*(x) \psi_l^*(x') H(x\hat{p}_x, x'\hat{p}_{x'}) \psi_m(x) \psi_n(x') \tag{2.43}$$

This can be proven by taking the matrix representation and showing that it is identical to the one obtained in the first quantization case (as illustrated by the simple example above). For a thorough discussion of this issue see ref. [3, 14].

Field operators $Y(x), Y^{\dagger}(x)$ are formed by linear combinations of creation and annihilation operators (denoted here $c^{\dagger}$ and c for generality) where the combination coefficients are the single-particle wave functions and the sum is over the complete set of single-particle quantum numbers.

$$Y(x) = \sum_{k} \psi_k(x) c_k \tag{2.44}$$

$$Y^+(x) = \sum_k \psi_k(x)^* c_k^\dagger \tag{2.45}$$

The field operators are operators in abstract occupation-number space because they depend on the creation and annihilation operators. The field operators satisfy commutation or anticommutation relations depending on the statistics. We note that $Y^\dagger$ creates a particle at position x and Y destroys a particle at x. The field operators satisfy the relation

$$[Y(x), Y^\dagger(x')] = \delta(x - x') \tag{2.46}$$

and the density operator (see Chapter 12) is given by $\rho(x) = Y^\dagger(x)Y(x)$. Furthermore the Hamiltonian operator can be written in terms of these field operators

$$H = \int dx \; Y^\dagger(x)T(x)Y(x) + \frac{1}{2} \int dx \int dx' \; Y^\dagger(x)Y^\dagger(x')V(x,x')Y(x')Y(x) \tag{2.47}$$

This expression is readily verified, since integration over the spatial coordinates produces the single-particle matrix elements of the kinetic energy and potential energy, leaving us with a sum of these matrix elements multiplied by the appropriate creation and annihilation operators. We note that Y and $Y^\dagger$ are not wave functions but operators. Consider for example a general one-body operator in the first quantization

$$J = \sum_i J(x_i) \tag{2.48}$$

The corresponding SQ operator is given by

$$J = \sum_{ts} \langle r| J |s\rangle c_r^\dagger c_s = \int dx \sum_{rs} \psi_r(x)^* J(x) \psi_s(x) c_r^\dagger c_s$$

$$= \int dx \; Y^\dagger(x)J(x)Y(x) \tag{2.49}$$

We shall later (in Chapters 15, 17, and 19) use the field operators when solving a number of dynamical problems.

The SQ formulation has been used extensively in many areas of physics and chemistry—in electronic structure calculations, energy transfer calculations, and photon statistics, to name a few. The reason for dealing with the SQ formalism in the present monograph is that we shall later use the formulation in connection with electron transfer theories (see Chapters 17 and 18) as well as in the following sections, where we consider various model cases, where a formulation in the SQ framework facilitates the solution of the TDSE.

EXERCISES

2.1. Prove that the two number operators, $n_r = a_r^\dagger a_r$ and $n_s = a_s^\dagger a_s$ commute.

2.2. Use the relations of Eqs. (2.31) and (2.32) to interchange the two particles in the states one and two of the state vector $|1, 1, 0, 0, \ldots\rangle$ with the operator $a_2^\dagger a_3 a_1^\dagger a_2 a_3^\dagger a_1$.

2.1   APPLICATIONS OF SECOND QUANTIZATION

The second quantization formulation offers a compact formulation of the dynamics of complicated systems and also often suggests simpler methods for the solution as compared with the first quantization formalism. We will illustrate this by considering a few examples, where the SQ method is used for solving exactly the TDSE for model problems. These model descriptions contain the correct physics of the problem and can therefore be used as either exact solutions for the approximate Hamiltonian, or as a starting point for a solution under the more accurate Hamiltonian, i.e., a zeroth order solution on which the exact solution can be expanded. Conventional quantum mechanical treatments utilize an expansion of the wave function in suitable basis sets. However such expansions will be impossible to use on large systems. The SQ formalism takes advantage of the fact that the number of relevant operators to be included is much smaller than the number of accessible states. Special cases are those where the operators form a closed set with respect to commutations. Here the TDSE can be solved algebraically. In the following we will consider such cases, namely the forced oscillator, energy transfer to polyatomic systems, and phonons (vibrational excitations) in a solid.

2.1.1   The Linearly Forced Harmonic Oscillator

The harmonic oscillator is an important model system, which can describe some of the main features in energy transfer mechanisms in polyatomic molecules and solids. The energy transfer comes about by a perturbation of the oscillator, either through interaction with an external field—electromagnetic interaction—or a perturbation by a collision. In the case of a linear forcing, the time-dependent Schrödinger equation is solvable either in the ordinary (i.e., first) quantization, or more conveniently through the second quantization concept. We notice that the Hamiltonian for the isolated harmonic oscillator (HO) is

$$H = \frac{1}{2m} p_r^2 + \frac{1}{2} k(r - r_e)^2 = E \qquad (2.50)$$

where m is the reduced mass, k the force constant, r the oscillator coordinate, r_e the equilibrium value, and p_r the conjugate momentum.

Introducing the mass-weighted momenta and coordinates

$$P = \frac{p_r}{\sqrt{m}} \qquad (2.51)$$

$$r - r_e = \frac{Q}{\sqrt{m}} \qquad (2.52)$$

we have

$$H = \frac{1}{2} (P^2 + \omega^2 Q^2) \qquad (2.53)$$

where the vibrational frequency ω is related to the force constant by $k = m\omega^2$. Introducing now the creation and annihilation operators $b^\dagger$ and b we write

$$\hat{Q} = \sqrt{\frac{\hbar}{2\omega}} (b^\dagger + b) \qquad (2.54)$$

and

$$\hat{P} = i\sqrt{\frac{\hbar\omega}{2}} (b^\dagger - b) \qquad (2.55)$$

This representation of the coordinate and momentum operators will give a Hamiltonian that generates the correct harmonic oscillator eigenvalues, matrix elements, and other properties known from first quantization. Thus introducing these expressions in the Hamiltonian we obtain

$$H_0 = \hbar\omega(b^\dagger b + \tfrac{1}{2}) \qquad (2.56)$$

We have in the derivation used the commutation relation

$$[b, b^\dagger] = 1 \qquad (2.57)$$

Furthermore

$$b|n\rangle = \sqrt{n} \, |n - 1\rangle \tag{2.58}$$

$$b^\dagger |n\rangle = \sqrt{n + 1} \, |n + 1\rangle \tag{2.59}$$

where n is the vibrational quantum number. Using these two relations we see that

$$H_0|n\rangle = \hbar\omega(n + \tfrac{1}{2})|n\rangle \tag{2.60}$$

i.e., we obtain the usual expression for the energy levels of the harmonic oscillator. The linearly forced harmonic oscillator problem is now obtained by perturbing the oscillator with the following time-dependent potential

$$V(t) = f(t)Q = A(b + b^\dagger)f(t) \tag{2.61}$$

where $A = \sqrt{\hbar/2\omega}$. The perturbation can be due to either an external time-dependent field from a laser, or a colliding particle, the motion of which is described classically, so as to obtain a time-dependent force.

In terms of creation/annihilation operators we write the total Hamiltonian as

$$H(t) = H_0 + V(t) = \hbar\omega(b^\dagger b + \tfrac{1}{2}) + Af(t)(b + b^\dagger) \tag{2.62}$$

We now wish to solve the time-dependent Schrödinger equation using an algebraic approach, i.e., the solution is sought within the operator space. Thus we have

$$i\hbar \, \frac{d\Psi(t)}{dt} = H(t)\Psi(t) \tag{2.63}$$

where we can initially have the oscillator in a given quantum state, i.e., in the coordinate representation

$$\Psi(t_0) = \phi_n = \langle r|n\rangle \tag{2.64}$$

We may now introduce the evolution operator $U(t, t_0)$ such that the wave function at time t is given as

$$\Psi(t) = U(t, t_0)\Psi(t_0) \tag{2.65}$$

Inserting this in the TDSE we see that $U(t, t_0)$ must satisfy the equation

$$ih\frac{dU(t, t_0)}{dt} = (H_0 + V(t))U(t, t_0) \tag{2.66}$$

It is often convenient to work in the so-called interaction representation, here we introduce $U_I(t, t_0)$ such that

$$U(t, t_0) = \exp(-iH_0 t/\hbar)U_I(t, t_0) \tag{2.67}$$

From the equation for $U(t, t_0)$ we then get

$$i\hbar\dot{U}_I = \exp(iH_0 t/\hbar)V(t)\,\exp(-iH_0 t/\hbar)U_I \tag{2.68}$$

Then inserting the operators we have

$$i\hbar\dot{U}_I = A f(t)\,\exp[i\omega t(b^\dagger b + \tfrac{1}{2})](b^\dagger + b)\,\exp[-i\omega t(b^\dagger b + \tfrac{1}{2})]U_I \tag{2.69}$$

By using the commutation relation in Eq. (2.57) it is possible to show that

$$\exp(iH_0 t/\hbar)b^\dagger = \sum_{n=0}^{\infty}\left(\frac{it}{\hbar}\right)^n \frac{1}{n!} H_0^n b^\dagger = b^\dagger\,\exp(i\omega t + iH_0 t/\hbar) \tag{2.70}$$

and that

$$\exp(iH_0 t/\hbar)b = b\,\exp(-i\omega t + iH_0 t/\hbar) \tag{2.71}$$

Thus we get

$$i\hbar\dot{U}_I = A f(t)[b^\dagger\,\exp(i\omega t) + b\,\exp(-i\omega t)]U_I \tag{2.72}$$

If we now use the Magnus expansion (see Appendix A) to solve Eq. (2.72) we find the solution of the TDSE to be

$$U(t, t_0) = \exp(-iH_0 t/\hbar)U_I(t, t_0) \tag{2.73}$$

where

$$U_I(t, t_0) = \exp[i(\beta(t, t_0) + \alpha^+(t, t_0)b^\dagger + \alpha^-(t, t_0)b)] \tag{2.74}$$

and

$$\alpha^{\pm}(t, t_0) = -\frac{A}{\hbar} \int_{t_0}^{t} dt' f(t') \exp(\pm i\omega t') \tag{2.75}$$

$$\beta(t, t_0) = \frac{A^2}{\hbar^2} \int_{t_0}^{t} dt' f(t') \int_{t_0}^{t'} dt'' f(t'') \sin[\omega(t' - t'')] \tag{2.76}$$

Since the evolution operator is now known in terms of the functions $\alpha^{\pm}$ and β, we can calculate the amplitude and the transition probability for the event that the system, which initially is in a given quantum state $|n\rangle$ at time $t = t_0$, ends up in state $|m\rangle$ at time t. Thus we have

$$\alpha_{mn}(t, t_0) = \langle m|U(t, t_0)|n\rangle \tag{2.77}$$

and for the probability

$$P_{n \to m}(t, t_0) = |\alpha_{mn}(t, t_0)|^2 \tag{2.78}$$

Using Eq. (2.73) we obtain

$$\alpha_{mn}(t, t_0) = \exp\left[-i\omega(m + \tfrac{1}{2})t\right]\langle m|U_I(t, t_0)|n\rangle \tag{2.79}$$

In order to evaluate the last factor we need to use the Baker-Hausdorff formula, which states that

$$\exp(A)\exp(B) = \exp(C) \tag{2.80}$$

$$C = A + B + \tfrac{1}{2}[A, B] + \tfrac{1}{12}[[A, B], B] + \tfrac{1}{12}[[B, A], A] + \ldots \tag{2.81}$$

where A, B, and C are operators. Using this formula we see that

$$\exp(i\alpha^+ b^\dagger + i\alpha^- b) = \exp\left(-\tfrac{1}{2}\rho(t, t_0)\right)\exp(i\alpha^+ b^\dagger)\exp(i\alpha^- b) \tag{2.82}$$

Thus we have

$$\alpha_{mn} = \exp\left[-i\omega(m + \tfrac{1}{2})t - \tfrac{1}{2}\rho(t, t_0) + i\beta(t, t_0)\right]\langle m|\exp(i\alpha^+ b^\dagger) \\ \cdot \exp(i\alpha^- b)|n\rangle \tag{2.83}$$

where $\rho = \alpha^+ \alpha^-$. The matrix element can be evaluated by expanding the exponential functions in power series and using the relations

$$b^q|n\rangle = \sqrt{n(n-1)\ldots(n-q+1)}\,|n-q\rangle$$

$$= \sqrt{\frac{n!}{(n-q)!}}\,|n-q\rangle \tag{2.84}$$

$$b^{\dagger\,p}|n\rangle = \sqrt{(n+1)(n+2)\ldots(n+p)}\,|n+p\rangle$$

$$= \sqrt{\frac{(n+p)!}{n!}}\,|n+p\rangle \tag{2.85}$$

allowing us to obtain

$$\alpha_{mn}(t,t_0) = \exp\left[-i\omega\left(m+\frac{1}{2}\right)t\right]\exp\left(-\frac{1}{2}\rho(t,t_0) + i\beta(t,t_0)\right)$$

$$\times \sqrt{n!m!}\,(i\alpha^+(t,t_0))^{m-n}\sum_{q=0}^{n}\frac{(-1)^q}{q!}\rho^q\frac{1}{(n-q)!}\frac{1}{(m-n+q)!}$$

$$\tag{2.86}$$

where terms containing negative factorials are zero. The expression for the transition amplitude is, of course, the same as the one obtained using other methods, e.g., the first quantization or the Feynman path integral method (see Chapter 4).

The energy transfer to the oscillator due to the perturbation with the force $f(t)$ can be calculated to be

$$\Delta E = \sum_{m}\hbar\omega(m-n)P_{nm} = \hbar\omega\rho \tag{2.87}$$

Note that the energy transfer to the harmonic oscillator is independent of the initial state (n) in the case of linear forcing. This observation will be discussed below, where we consider energy transfer to solids.

It is possible to write the above expression in a more compact fashion by introducing the Lagrange polynomials L_n^α, i.e., for $m \geq n$

$$\alpha_{mn}(t,t_0) = \exp\left[-i\omega(m+\tfrac{1}{2})t\right]\exp\left(-\tfrac{1}{2}\rho(t,t_0) + i\beta(t,t_0)\right)$$

$$\cdot (i\alpha^+(t,t_0))^{m-n}\sqrt{n!/m!}\,L_n^{m-n}(\rho) \tag{2.88}$$

Often we will work with large quantum numbers. In this limit, further simplification can be introduced. Thus in the limit of large values of n and m we have

$$P_{nm} = |\alpha_{mn}|^2 \sim \exp(-\rho)J^2_{m-n}(2\sqrt{n\rho}) \tag{2.89}$$

where J_n is a Bessel function.

The linearly forced harmonic oscillator is one example where the TDSE can be solved analytically using second quantization methods; but other examples exist, e.g., the time-dependently forced Morse oscillator is also solvable [5] and, in general, problems where the Hamiltonian operator can be factorized [6] are candidates for an exact solution of the TDSE. However, in general we must require that the operators form a closed set with respect to commutations. We shall illustrate this further by considering energy transfer to polyatomic molecules in Section 2.1.3.

EXERCISES

2.3. Derive Eq. (2.70).

2.4. Derive Eq. (2.87) for $n = 0$.

2.5. Find the energy transfer to a classical harmonic oscillator by solving the equations of motion obtained with the Hamiltonian

$$H = \frac{1}{2} P^2 + \frac{\omega^2}{2} Q^2 + f(t)Q \tag{2.90}$$

2.1.2 Coherent States

In the quantum theory of light the so-called coherent states (see also Chapter 18) play an important role. An important relationship for this discussion is the Heisenberg uncertainty relation

$$\Delta p \Delta q \geq \tfrac{1}{2}\hbar \tag{2.91}$$

where

$$\Delta p = \sqrt{\langle p^2 \rangle - \langle p \rangle^2} \tag{2.92}$$

is the root-mean-square deviation of p; a similar equation holds for q. The commutator relationship is

$$[p, q] = \frac{\hbar}{2} \tag{2.93}$$

The set of coherent states is a class of simple harmonic oscillator states that minimize the Heisenberg uncertainty relation and simultaneously minimize the spread in both position (q) and momentum operators (p). Another option is to minimize the uncertainty product such that Δp is much smaller than Δq or vice versa, this leads to squeezed coherent states.

From Section 2.1 we have that

$$b = \frac{ip}{\sqrt{2\hbar\omega m}} + \sqrt{\frac{k}{2\hbar\omega}}\, q \tag{2.94}$$

and

$$b^\dagger = \frac{-ip}{\sqrt{2\hbar\omega m}} + \sqrt{\frac{k}{2\hbar\omega}}\, q \tag{2.95}$$

For the harmonic oscillator the equations of motion are

$$\frac{dx}{dt} = \frac{p(t)}{m} \tag{2.96}$$

and

$$\frac{dp}{dt} = -kx(t) \tag{2.97}$$

which in the Heisenberg picture give the following solutions

$$q(t) = q(0)\cos\omega t + \frac{p(0)}{\omega m}\sin\omega t \tag{2.98}$$

$$p(t) = -\omega m q(0)\sin\omega t + p(0)\cos\omega t \tag{2.99}$$

The expectation values of $p(t)$ and $q(t)$ for a given state $|\psi\rangle$ are evaluated as

$$\langle p(t)\rangle = \langle\psi|p(t)|\psi\rangle \tag{2.100}$$

$$\langle q(t)\rangle = \langle\psi|q(t)|\psi\rangle \tag{2.101}$$

In terms of the creation and destruction operators we have the solutions

$$b(t) = \exp(-i\omega t)b(0) \tag{2.102}$$

$$b^\dagger(t) = \exp(i\omega t)b^\dagger(0) \tag{2.103}$$

In order to introduce the coherent state we start out by defining

$$G(t) = b(t) - \langle b(t) \rangle \tag{2.104}$$

$$G^\dagger(t) = b^\dagger(t) - \langle b^\dagger(t) \rangle \tag{2.105}$$

which fulfill the same commutation relationships as the creation and annihilation operators $b^\dagger, b$. We are able to write the variances of $p(t)$ and $q(t)$ as

$$\langle \Delta q(t)^2 \rangle = \frac{\hbar}{2\sqrt{km}} \left[\langle G(0)^2 \rangle \exp(-2i\omega t) + \langle G^\dagger(0)^2 \rangle \exp(2i\omega t) \right.$$

$$\left. + \langle G^\dagger(0)G(0) + G(0)G^\dagger(0) \rangle \right] \tag{2.106}$$

and

$$\langle \Delta p(t) \rangle^2 = \frac{\sqrt{km}\hbar}{2} \left[\langle G(0)^2 \rangle \exp(-2i\omega t) + \langle G^\dagger(0)^2 \rangle \exp(2i\omega t) \right.$$

$$\left. - \langle G^\dagger(0)G(0) + G(0)G^\dagger(0) \rangle \right] \tag{2.107}$$

The correlation between $p(t)$ and $q(t)$ is given by

$$\left\langle \frac{1}{2} [\Delta p(t)\Delta q(t) + \Delta q(t)\Delta p(t)] \right\rangle$$

$$= -\frac{i\hbar}{2} [\langle G(0)^2 \rangle \exp(-2i\omega t) - \langle G^\dagger(0)^2 \rangle \exp(2i\omega t)] \tag{2.108}$$

The variances of $p(t)$ and $q(t)$ will be time independent when both $\langle G(0)^2 \rangle$ and $\langle G^\dagger(0)^2 \rangle$ are zero. This corresponds to the situation where the wavepacket moves with constant uncertainties in $p(t)$ and $q(t)$.

We obtain the minimum uncertainty by minimizing $\langle G^\dagger(0)G(0)+G(0)G^\dagger(0) \rangle$. Since the operators $G^\dagger(t)$ and $G(t)$ obey the same commutation relations as $b^\dagger(t)$ and $b(t)$ we find that the eigenvalues of $G^\dagger(0)G(0)$ will be the same as the ones for the operator $b^\dagger b$. Thus the eigenvalues of $G^\dagger(0)G(0) + G(0)G^\dagger(0)$ are of the form $n + \frac{1}{2}$ where $n \geq 0$. The minimum value of $\langle G^\dagger(0)G(0)+G(0)G^\dagger(0) \rangle$ is for $n = 0$ and we call this state $|\alpha\rangle$. We have

$$G(0)|\alpha\rangle = 0 \tag{2.109}$$

The coherent states are characterized by being eigenfunctions to the photon annihilation operator. Introducing the notation from Section 2.1.1, Eqs. (2.58) and (2.59), we then have

$$b\Phi_- = \alpha^-\Phi_- \qquad (2.110)$$

where Φ_- is a coherent state. The states are formed from the vacuum state as

$$\Phi_- = \exp(-\rho/2)\,\exp(\alpha^- b^\dagger)|vac\rangle \qquad (2.111)$$

where $\rho = \alpha^+\alpha^-$ and can be obtained by taking a Poisson distribution of n-photon states, i.e.,

$$\Phi_- = \sum_n c_n\Phi_n \qquad (2.112)$$

where $\Phi_0 = |vac\rangle$,

$$c_n = \frac{(\alpha^-)^n}{\sqrt{n!}}\,\exp(-\rho/2) \qquad (2.113)$$

and the n-photon state

$$\Phi_n = \frac{1}{\sqrt{n!}}\,(b^\dagger)^n\Phi_0 \qquad (2.114)$$

An operator that occurs often in quantum optics and quantum mechanics is considered in Exercise 2.7 below.

EXERCISES

2.6. Derive Eq. (2.111).

2.7. Show that the coherent state is an eigenfunction to the displaced harmonic oscillator, i.e., to $\hbar\omega b^\dagger b + \gamma^* b^\dagger + \gamma b$.

2.1.3 Energy Transfer in Polyatomic Molecules

In large polyatomic molecules the most important degrees of freedom are the $3N - 6$ vibrational modes, and energy transfer to these modes constitutes the dominating energy loss mechanism in collisions. In the section above, we saw how energy transfer to a single harmonic oscillator could be described within the second quantization formalism. In the present section, we will show that the SQ method can also be used for solution of the TDSE with a general Hamiltonian expressed as

$$H = \sum_k \hbar\omega_k \left(b_k^\dagger b_k + \frac{1}{2} \right) + \sum_k F_k(t)(b_k + b_k^\dagger) + \sum_{kl} F_{kl}(t)b_k^\dagger b_l \quad (2.115)$$

This Hamiltonian is a model Hamiltonian for a polyatomic molecule perturbed by linear and quadratic forces

$$F_k = \frac{\partial V_{int}}{\partial Q_k} \bigg|_{eq} \quad (2.116)$$

$$F_{kl} = \frac{1}{2} \frac{\partial^2 V_{int}}{\partial Q_k \partial Q_l} \bigg|_{eq} \quad (2.117)$$

where V_{int} is the intermolecular potential and the coordinates Q_k are normal-mode coordinates. If a classical path is assumed for the relative motion of two colliding molecules, the forces become time dependent through the classical equations of motion (the "classical path method" [7]). However the forcing can also arise as a consequence of an external field, and the model Hamiltonian then describes field–molecule interaction in the semiclassical limit, where the field is treated classically.

We now consider the time-dependent Schrödinger equation, i.e.,

$$i\hbar \frac{d\Psi(t)}{dt} = H\Psi(t) \quad (2.118)$$

Introducing the evolution operator in the interaction representation, i.e.,

$$U(t, t_0) = \exp(-iH_0 t/\hbar)U_I(t, t_0) \quad (2.119)$$

where $H_0 = \sum_k \hbar\omega_k(b_k^\dagger b_k + \frac{1}{2})$ we obtain

$$i\hbar \frac{dU_I}{dt} = \left[\sum_k (f_k^+ b_k^\dagger + f_k^- b_k) + \sum_{kl} (f_{kl}^+ b_k^\dagger b_l + f_{kl}^- b_k b_l^\dagger) \right] U_I \quad (2.120)$$

where

$$f_k^{\pm} = F_k \ \exp(\pm i\omega_k t) \tag{2.121}$$

$$f_{kl}^{\pm} = F_{kl} \ \exp[\pm i(\omega_k - \omega_l)t] \tag{2.122}$$

Considering the solution to the equation including just the quadratic operator terms, i.e.

$$i\hbar \frac{dU_{VV}}{dt} = \sum_{kl} (f_{kl}^+ b_k^\dagger b_l + f_{kl}^- b_k b_l^\dagger)U_{VV} \tag{2.123}$$

we can write

$$U_I = U_{VV}U_{VT} \tag{2.124}$$

where U_{VT} can be factorized as $U_{VT} = \Pi_k U_{VT}^{(k)}$ and $U_{VT}^{(k)}$ must satisfy the equation

$$i\hbar \frac{dU_{VT}^{(k)}}{dt} = U_{VV}^\dagger(f_k^+ b_k^\dagger + f_k^- b_k)U_{VV}U_{VT}^{(k)} \tag{2.125}$$

The solution U_{VV} to Eq. (2.123) can be expressed in terms of the operators $O_{kl} = b_k^\dagger b_l$, i.e.,

$$U_{VV} = \Pi_{ij} \ \exp(\alpha_{ij}(t,t_0)O_{ij}) \tag{2.126}$$

The reason for this is that the operators O_{ij} form a closed set with respect to commutations. The as-yet-unknown functions $\alpha_{ij}(t,t_0)$ are related to the solutions of the coupled equations

$$i\hbar \frac{d\mathbf{R}(t,t_0)}{dt} = \mathbf{BR}(t,t_0) \tag{2.127}$$

where $\mathbf{R}(t_0,t_0) = \mathbf{I}$, the unit matrix. The elements of the B-matrix are

$$B_{kl} = f_{kl}^+ (k > l) \tag{2.128}$$

$$B_{kl} = f_{kl}^- (k < l) \tag{2.129}$$

i.e., of dimension $M \times M$ where M^2 is the number of operators O_{kl} entering the problem. In order to solve the problem posed by Eqs. (2.125) we have to evaluate the quantities

$$U_{VV}^\dagger(b_k^\dagger \ \text{or} \ b_k)U_{VV} \tag{2.130}$$

This can be done using operator algebra. The result is [8]

$$U_{VV}^{\dagger} b_k U_{VV} = b_k + \sum_{j=1}^{M} Q_{kj} b_j \qquad (2.131)$$

$$U_{VV}^{\dagger} b_k^{\dagger} U_{VV} = b_k^{\dagger} + \sum_{j=1}^{M} Q_{kj}^{*} b_j^{\dagger} \qquad (2.132)$$

where $Q_{kj} = R_{kj} - \delta_{kj}$. Thus the "VT" problem can be transformed to [9]

$$i\hbar \frac{d U_{VT}^{(k)}}{dt} = (W_k^{+} b_k^{\dagger} + W_k^{-} b_k) U_{VT}^{(k)} \qquad (2.133)$$

where

$$W_k^{+} = f_k^{+} + \sum_{j=1}^{M} f_j^{+} Q_{jk}^{*} \qquad (2.134)$$

and $W_k^{-} = (W_k^{+})^{*}$. Equation (2.133) can be solved using the algebraic method given in the previous section. The terms "VV" and "VT" come from molecular collision theory [7]. The VV-operators induce transition between two vibrational modes, such that the number of vibrational quanta are conserved. For the remaining processes we use the term VT-processes. The VT-operators excite a specific mode through vibrational–translational coupling.

The operator approach for the vibrational degrees of freedom is conveniently used in the classical path method, where the relative translational motion of the two molecules is treated classically, i.e., the classical equations of motion are solved simultaneously with the Eqs. (2.127 (for the VV-problem) and the calculation of the integrals Eqs. (2.75) and (2.76) (for the VT-problem). The quantum and classical subsystems can be coupled self-consistently (see also Chapter 3) through an effective Hamiltonian written as

$$H_{\text{eff}} = \langle \Psi_0 | U_{VT} U_{VV} | H | U_{VV} U_{VT} | \Psi_0 \rangle \qquad (2.135)$$

where Ψ_0 is the initial wave function, i.e.,

$$\Psi(t) = \exp(-iH_0 t)/\hbar) U_I(t, t_0) \Psi_0 \qquad (2.136)$$

and the integration is over the "quantum coordinates." Thus the effective Hamiltonian H_{eff} will depend upon time either explicitly or through the classical coordinates and momenta. The effective Hamiltonian can be expressed in terms of the functions $\alpha_k^{\pm}$ and Q_{kj}, using the Eqs. (2.131) and (2.132) and

$$b_k \, \exp(i\alpha_k^+ b_k^\dagger) = \exp(i\alpha_k^+ b_k^\dagger)(b_k + i\alpha_k^+) \tag{2.137}$$

$$b_k^\dagger \, \exp(i\alpha_k^- b_k) = \exp(i\alpha_k^- b_k)(b_k^\dagger - i\alpha_k^-) \tag{2.138}$$

The equations of motion for the classical degrees of freedom are obtained using Hamilton's principle on H_{eff}, i.e.

$$\frac{dH_{\mathrm{eff}}}{dt} = \frac{\partial H_{\mathrm{eff}}}{\partial t} + \frac{\partial H_{\mathrm{eff}}}{\partial R}\dot{R} + \frac{\partial H_{\mathrm{eff}}}{\partial P_R}\dot{P}_R = 0 \tag{2.139}$$

Unless the operators U_{VV} and U_{VT} depend explicitly upon some of the classical variables then $\partial H_{\mathrm{eff}}/\partial t = 0$ and we have

$$\dot{R} = \frac{\partial H_{\mathrm{eff}}}{\partial P_R} \tag{2.140}$$

$$\dot{P}_R = -\frac{\partial H_{\mathrm{eff}}}{\partial R} \tag{2.141}$$

The methodology presented above is extremely powerful for treating energy transfer in polyatomic molecules, retaining a quantum description of the vibrational modes (see e.g., ref. [10]).

In order to obtain the solution we have made use of the fact that the time-dependent Schrödinger equation is algebraically solvable if the operators in the Hamiltonian form a closed set with respect to commutations. The closed set of operators constitute a Lie group and hence group theoretical methods can be used (see e.g. [11]) when solving the equations.

The expression for the transition amplitude from a state $|n_1 \ldots n_M\rangle$ to $|n_1' \ldots n_M'\rangle$ induced by the operator U_{VV} is given by [7]

$$\langle n_M' \ldots n_1' | U_{VV} | n_1 \ldots n_M \rangle$$

$$= \exp\left(\sum_{j=1}^M \alpha_{jj}\right) \times \sqrt{\frac{n_M! \Pi_{j=2}^M n_j'!}{n_1'! \Pi_{j=1}^{M-1} n_j!}} \sum_{k_{ji}=0, \, i \neq j}^{\infty} \Pi_{i \neq j} \left(\frac{\alpha_{ji}^{k_{ji}}}{k_{ji}!}\right)$$

$$\cdot \frac{\Pi_{p=1}^{M-1}\left(n_p + \sum_{q=p+1}^M k_{pq}\right)!}{\Pi_{p=2}^M\left(n_p + \sum_{q=p+1}^M k_{pq} - \sum_{q \neq p} k_{qp}\right)!} \tag{2.142}$$

where the time-dependent functions α_{ji} are related to the solution of Eqs. (2.127) (see [7] and Appendix B). The VT-transition amplitudes for transitions induced by the operators $U_{VT}^{(k)}$ can be obtained using Eq. (2.86). Thus the numerical effort is reduced to the solution of the coupled Eqs. (2.127), M integrals of the type in Eqs. (2.75) and (2.76), and equations of motion for the classical variables.

EXERCISE

2.8. Show that $\partial H_{\text{eff}}/\partial t = 0$ for a quantum classical Hamiltonian that depends upon time only through the classical variables, i.e.,

$$H = H_{\text{cl}}(P_R(t), R(t)) + H_q(R(t); \{b_k, b_k^\dagger\}) \tag{2.143}$$

where we have indicated that the quantum part depends upon the operators b_k and $b_k^\dagger$.

2.1.4 Energy Transfer to Solids

Energy transfer to solids is conveniently treated within the second quantization approach. The quantized vibrational modes of the solid (phonons) are bosons, i.e., we can introduce the boson operators defined previously. In order to apply the SQ scheme we introduce the normal mode coordinates of the solid. The transformation between displacement ($\eta_{\alpha\gamma}$) and normal mode (Q_r) coordinates is

$$\eta_{\alpha\gamma} = \sum_r T_{\alpha\gamma;r} Q_r \tag{2.144}$$

where α denotes an atom in the solid, γ takes the values 1, 2, 3 denoting the x, y, z directions and Q_r is the rth normal mode. $T_{\alpha\gamma;r}$ are elements of the **T** matrix defined such that it diagonalizes the force constant matrix **A** of the solid, with elements given by

$$A_{k\gamma;l\delta} = \frac{1}{\sqrt{M_k M_l}} \frac{\partial^2 V}{\partial r_{k\gamma} \partial r_{l\delta}} \tag{2.145}$$

where $k, l = 1, \ldots, N$ and $\gamma, \delta = 1, 2, 3$. Thus the Hamiltonian (H_c) for the solid includes first and second order terms in the displacement from the equilibrium positions

$$H_c = V_0 + \sum_{r=1}^{M} H_r \tag{2.146}$$

where V_0 is the zero point energy

$$H_r = \tfrac{1}{2}(\dot{Q}_r^2 + \omega_r^2 Q_r^2) \qquad (2.147)$$

The $M = 3N - 6$ non-zero frequencies are obtained by diagonalizing the second derivative matrix. We can now quantize the vibrational modes r and write the normal mode coordinates in terms of the boson creation/annihilation operators

$$Q_r = A_r(b_r^\dagger + b_r) \qquad (2.148)$$

where $A_r = (\hbar/2\omega_r)^{1/2}$.

The probability for oscillator k to be in quantum state n_k is usually taken as a Boltzmann distribution, i.e.,

$$P_{n_k} = z_k^{n_k}(1 - z_k) \qquad (2.149)$$

where

$$z_k = \exp(-\beta\hbar\omega_k) \qquad (2.150)$$

$\beta = 1/kT_s$, and T_s is the temperature of the solid.

Transitions among the phonon quantum states may now be induced by anharmonic terms or through coupling to the electronic motion. Transitions induced by gas–surface collisions are, however, the processes that are the concern of the present section. We introduce $R_{a\alpha}$ as the distance from the gas phase atom a to the surface atom α, i.e.,

$$R_{a\alpha} = \sqrt{\sum_{i=1,2,3}(\overline{X}_i - X_{\alpha i})^2} \qquad (2.151)$$

with the gas atom positions given by $(\overline{X}_1, \overline{X}_2, \overline{X}_3)$ and the positions $X_{\alpha i}$ of the solid atoms. We can now expand the interaction potential in normal mode coordinates as

$$V_I = V_I^{(0)} + \sum_r V_r^{(1)} Q_r + \frac{1}{2} \sum_{rr'} V_{rr'}^{(2)} Q_r Q_{r'} \qquad (2.152)$$

where $V_I^{(0)}$ is the interaction potential with the lattice atoms at their equilibrium position $X_{\alpha i}^{eq}$ and

$$V_r^{(1)} = \sum_\alpha \left. \frac{\partial V_{a\alpha}}{\partial R_{a\alpha}} \right|_{eq} \frac{\partial R_{a\alpha}}{\partial Q_r} \tag{2.153}$$

is the first derivative of the interaction potential with respect to phonon coordinate Q_r. The second derivative is denoted by $V_{rr'}^{(2)}$ and given by

$$V_{rr'}^{(2)} = \sum_\alpha \left. \frac{\partial V_{a\alpha}}{\partial R_{a\alpha}} \right|_{eq} \frac{\partial^2 R_{a\alpha}}{\partial Q_r \partial Q_{r'}} + \sum_{\alpha\beta} \left. \frac{\partial^2 V_{a\alpha}}{\partial R_{a\alpha} \partial R_{a\beta}} \right|_{eq} \frac{\partial R_{a\alpha}}{\partial Q_r} \frac{\partial R_{a\beta}}{\partial Q_{r'}} \tag{2.154}$$

In order to calculate $\partial R_{a\alpha}/\partial Q_r$ we use Eqs. (2.144) and (2.151), and that

$$\eta_{\alpha\gamma} = m_\alpha^{1/2}(X_{\alpha\gamma} - X_{\alpha\gamma}^{eq}) \tag{2.155}$$

i.e.,

$$\frac{\partial R_{a\alpha}}{\partial Q_r} = -m_\alpha^{-1/2} \sum_{i=1}^{3} T_{\alpha i ;r}(\overline{X}_i - X_{\alpha i}^{eq})/R_{a\alpha}^{eq} \tag{2.156}$$

The second derivatives can be expressed as

$$\frac{\partial^2 R_{a\alpha}}{\partial Q_r \partial Q_{r'}} = \frac{1}{m_\alpha R_{a\alpha}^3} \sum_{ij} T_{\alpha i ;r} T_{\alpha j ;r'} (\overline{X}_i - X_{\alpha i}^{eq})(\overline{X}_j - X_{\alpha j}^{eq})(\delta_{ij} - 1) \tag{2.157}$$

The terms $V_r^{(1)}$ and $V_{rr'}^{(2)}$ in Eq. (2.152) are those responsible for collision-induced phonon excitation. These terms depend on time through the time-dependence of the trajectory. If only the linear terms $V_r^{(1)}$ are included, the oscillators are denoted "linearly forced." Omitting the zero-point energy of the solid (V_0) we can write the total Hamiltonian for an atom colliding with the surface as [8]

$$H = \frac{1}{2} \sum_{r=1}^{M} (\dot{Q}_r^2 + \omega_r^2 Q_r^2) + V_I^{(0)} + \sum_r V_r^{(1)} Q_r$$

$$+ \frac{1}{2} \sum_{rr'} V_{rr'}^{(2)} Q_r Q_{r'} + \frac{1}{2M} P_R^2 \tag{2.158}$$

where the last term is the kinetic energy of the atom with mass M. The evolution operator for the phonons perturbed linearly and quadratically is, as mentioned above, known from the theory for collision-induced energy transfer in polyatomic molecules. It may be expressed in terms of the boson creation/annihilation operators. In the classical path treatment the dynamics of the gas-phase atoms is treated classically. The effective Hamiltonian that governs this motion is then conveniently approximated by

$$H_{\text{eff}} = \langle \Psi|H|\Psi \rangle = \langle \Psi_0|U^\dagger H U|\Psi_0 \rangle \tag{2.159}$$

where $U(t, t_0)$ is the evolution operator and $|\Psi_0\rangle$ the initial wave function, i.e.,

$$|\Psi_0\rangle = \Pi^M_{k=1}\, |n^0_k\rangle \tag{2.160}$$

If only linear terms in Q_{lk} and a^+_k (see previous section) are included, the final expression for the effective Hamiltonian is [8]

$$H_{\text{eff}} = \sum_k \hbar\omega_k(\rho_k + n^0_k) + V^{(0)}_I + \sum_k \omega^{-1}_k\epsilon_k(t) + \frac{P^2_R}{2M}$$

$$+ \frac{1}{2} \sum_{kl} A_k A_l V^{(2)}_{kl}\{(2n^0_l + 1)\delta_{lk} + \exp[i(\omega_k - \omega_l)t](n^0_k Q_{lk} + n^0_l Q^*_{kl})$$

$$+ \exp[i(\omega_l - \omega_k)t](Q^*_{lk}(n^0_k + 1) + Q_{kl}(n^0_l + 1)]\} + O(Q^2_{kl}) \tag{2.161}$$

In the derivation we have used that the initial wave function is in a "pure" state and that

$$U^\dagger_{VT}U^\dagger_{VV}b_k U_{VV}U_{VT} = b_k + i\alpha^+_k + \sum_j Q_{kj}(b_j + i\alpha^+_j) \tag{2.162}$$

$$U^\dagger_{VT}U^\dagger_{VV}b^\dagger_k U_{VV}U_{VT} = b^\dagger_k - i\alpha^-_k + \sum_j Q^*_{kj}(b^\dagger_j - i\alpha^-_j) \tag{2.163}$$

We have furthermore neglected terms of second or higher order in Q_{kl} and α^+_k. The functions $Q_{lk}(t)$ are, to first order given, by

$$Q_{kl}(t) = Q_{lk}(t) = \frac{A_k A_l}{2i\hbar} \int^t_{t_0} dt' V^{(2)}_{kl}(t')\, \exp[i(\omega_k - \omega_l)t] \tag{2.164}$$

where the trajectory for the incoming atom is assumed to start at $t = t_0$. In Eq. (2.161) ρ_k is the so-called "excitation strength" for oscillator k, i.e., a measure of how much this particular phonon mode is excited. Initially we then have $\rho_k(t_0) = 0$, and finally, i.e., for $t \to \infty$ where the atom has left the surface, the only remaining terms in Eq. (2.161) are

$$E_{\text{int}} = \sum_{k=1}^{M} \hbar\omega_k \rho_k \tag{2.165}$$

and the kinetic energy term. Thus E_{int} is the energy transfer to the solid. The quantity $\epsilon_k(t)$ in Eq. (2.161) is defined as

$$\epsilon_k(t) = V_k^{(1)}(R(t)) \int_{t_0}^{t} dt' V_k^{(1)}(R(t')) \sin[\omega_k(t' - t)] \tag{2.166}$$

This term arises from the expectation value of the linear $V_k^{(1)} Q_k$ term in Eq. (2.158). These terms are called phonon VT terms because they exchange one quantum of vibrational energy when forced by the translational motion of the incoming atom. The second derivative terms in Eq. (2.158) contain the operators $b_r^{\dagger} b_{r'}$ and $b_r b_{r'}^{\dagger}$, i.e., these terms create a quantum in one mode and destroy one in another. It turns out, however, that the so-called VV terms are of minor importance for energy transfer to solids. These terms may therefore be neglected [23].

Equation (2.165) shows that $E_{\text{int}} > 0$, irrespective of the initial state and thereby of the surface temperature. The reason for this artifact is not the harmonic approximation but rather the self-consistent field (SCF) approximation (see Chapter 3) made when coupling the two subsystems, the atom and the solid. In order to improve this situation, corrections must be introduced (see Chapter 3, Section 3.3: Corrections to the SCF scheme).

If we were to neglect the VV processes we would get the following effective Hamiltonian from Eq. (2.161)

$$H_{\text{eff}} = E_{\text{int}} + V_I^{(0)}(R) + \sum_k \omega_k^{-1} \epsilon_k(R, t) + \frac{P_R^2}{2M} \sum_k \hbar\omega_k n_k^0 \tag{2.167}$$

along with the equations of motion, are obtained using Hamilton's principle, i.e. that,

$$\frac{dH_{\text{eff}}}{dt} = \frac{\partial H_{\text{eff}}}{\partial t} + \frac{\partial H_{\text{eff}}}{\partial R} \dot{R} + \frac{\partial H_{\text{eff}}}{\partial P_R} \dot{P_R} \tag{2.168}$$

yield

$$\dot{R} = \frac{\partial H_{\text{eff}}}{\partial P_R} = \frac{P_R}{M} \tag{2.169}$$

and

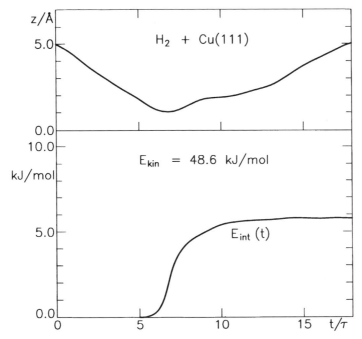

Fig. 2.1. Energy transfer to a Cu(111) surface by collision with a hydrogen molecule. The initial kinetic energy is 48.6 kJ/mole and vibrational rotational states are $(v,j) =$ (0, 0). The surface temperature is 300 K and incident angles $(\theta,\phi) = (0, 0)$.

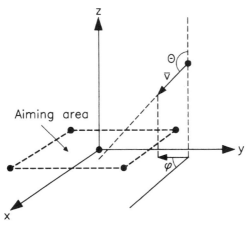

Fig. 2.2. Aiming area, incident angles Θ and ϕ, initial velocity vector v, and its projection on the surface for atom-surface scattering.

$$\dot{P}_R = -\frac{\partial H_{\text{eff}}}{\partial R} = -\frac{\partial V_I^{(0)}}{\partial R} + \sum_k \omega_k^{-1} \frac{\partial \epsilon_k(R, t)}{\partial R} \tag{2.170}$$

Thus if we require $\dot{H}_{\text{eff}} = 0$ we get

$$\frac{\partial H_{\text{eff}}}{\partial t} = \sum_{k=1}^{M} \hbar \omega_k \dot{\rho}_k + \sum_k \omega_k^{-1} \frac{\partial \epsilon_k}{\partial t} = 0 \tag{2.171}$$

or

$$\frac{\partial \epsilon_k}{\partial t} = -\hbar \omega_k^2 \dot{\rho}_k = -\omega_k \frac{d}{dt} \Delta E_k \tag{2.172}$$

where $\Delta E_k = \hbar \omega_k \rho_k$. Equation (2.172) may be used to evaluate ϵ_k in the effective potential

$$\langle V^{(1)} \rangle = \sum_k \omega_k^{-1} \epsilon_k(R, t) \tag{2.173}$$

from ΔE_k. This observation has been used to incorporate a simple correction to the SCF-treatment of atom/molecule surface scattering and hence for obtaining an initial state dependence of the effective potential and the energy transfer [22]. This is essential for making it possible to include surface temperature effects on the energy transfer.

The electron-hole pair excitation, where the collision induces electronic transitions of electrons from below to above the Fermi level, may also be important for a realistic description of the energy loss to the solid. It is, therefore, important that these processes also can be described within the SQ approach, involving fermion operators rather than boson operators [22]. Figure 2.1 shows the energy transfer as a function of time for hydrogen colliding with a Copper-surface. The energy transfer is found by solving the above equations using a realistic potential for the hydrogen–copper interaction [12]. The energy transfer depends strongly upon the surface temperature and other characteristics such as surface corrugation, steps and anomalies, the aiming site, and incident angles Θ and ϕ (see Fig. 2.2), where Θ is the angle between the velocity vector and the surface normal and ϕ the angle between the X-axis and the surface projected velocity vector.

EXERCISE

2.9. Verify that $\int dt\, \partial H_{\text{eff}}/\partial t = 0$.

3

EFFECTIVE HAMILTONIAN APPROACHES

A reduction in the complexity of the dynamics of many-body systems can be achieved by working with approximate interactions. One can imagine that the system is divided in parts and that the interaction between the parts is described approximately. Most often we are interested in full dynamical resolution only of part of the system. One reason for this could be that experimental data are restricted, in the sense that they only probe part of the system. It is advantageous to utilize this fact when formulating the theory. Thus the dynamics of the subsystem of interest is described to full order, while that of the larger subsystem only approximately. The interaction between the two subsystems is then introduced through an effective Hamiltonian. Various theoretical approaches have been developed for the evaluation of this effective interaction.

In the time-dependent self-consistent field (TDSCF) approach, the solution to the TDSE is sought by expressing the wavefunction in a Hartree product type form (for example see ref. [13]). Considering just a two-dimensional system we then have

$$\Psi = \tilde{\phi}_1(x_1, t)\tilde{\phi}_2(x_2, t) \tag{3.1}$$

where the functions $\tilde{\phi}_i$ are assumed to form an orthonormal set. The Hamiltonian can be expressed as

$$H = H_1(x_1) + H_2(x_2) + V(x_1, x_2) \tag{3.2}$$

Upon inserting the above trial function in the TDSE we find

35

$$i\hbar \, \frac{\partial \phi_i(x_i, t)}{\partial t} = H_i^{SCF}(x_i, t)\phi_i(x_i, t) \tag{3.3}$$

where we have multiplied from the left with $\Psi/\tilde{\phi}_i(x_i, t)$, integrated over all coordinates other than x_i and introduced

$$\phi_i(x_i, t) = \tilde{\phi}_i(x_i, t) \exp\left(\int \sum_{j \neq i} \frac{i}{\hbar} \langle \tilde{\phi}_j | H_j | \tilde{\phi}_j \rangle \, dt - \int \langle \tilde{\phi}_j | \dot{\tilde{\phi}}_j \rangle \, dt \right) \tag{3.4}$$

where the brackets $\langle \rangle$ indicate integration over the coordinate x_j. We note that this result can be generalized to any dimension of the problem, if the wave function is taken in the simple product form in Eq. (3.1). The multidimensional problem has been reduced to a number of one-dimensional ones, the reason being that the SCF Hamiltonian is given as

$$H_i^{SCF}(x_i, t) = H_i(x_i) + \sum_{j \neq i} \langle \phi_j | V(x_i, x_j) | \phi_j \rangle \tag{3.5}$$

where as above the brackets $\langle \rangle$ indicate integration over the coordinate x_j, i.e., the Hamiltonian depends only upon x_i. If the potential

$$V^{SCF}(x_i) = \sum_{j \neq i} \langle \phi_j | V(x_i, x_j) | \phi_j \rangle \tag{3.6}$$

is localized or slowly varying in x_i we can introduce a semiclassical solution to Eq. (3.3) by introducing time-dependent Gaussian wavepackets.

3.1 GAUSSIAN WAVEPACKETS

Localized wavepackets can be constructed by taking a superposition of plane waves, i.e., as

$$\phi(x, t = 0) = \sum_k c_k \exp(ikx) \tag{3.7}$$

Through a suitable choice of the coefficients c_k and by transforming the summation to an integral we can obtain a Gaussian wavepacket. From this derivation it is obvious that the wavepacket covers a range (actually an infinite range) in momentum (or energy) space. Thus if energy-resolved quantities are needed one must make a time-energy or momentum-coordinate Fourier transform (see also Chapter 5). What makes the localized wavepackets so beneficial is that they resemble "classical" particles and, as we shall see below the motion of the center of the wavepacket follows classical mechanical equations of motion.

The application of Gaussian wavepackets (GWP) in dynamical processes has been pioneered mainly by Heller and coworkers [15]. Heller introduced the following notation

$$\phi_i(x_i, t) = \exp\left\{\frac{i}{\hbar}\left[(\gamma_i(t) + p_i(t)(x_i - x_i(t)) + A_i(t)(x_i - x_i(t))^2\right]\right\} \quad (3.8)$$

where $\gamma_i(t)$ is a phase factor, $p_i(t)$ a momentum, and $A_i(t)$ a width parameter. Inserting the above expression in Eq. (3.3), expanding the potential to second order around the path $x_i(t)$, and equating equal powers of $(x_i - x_i(t))^k$ we obtain

$$\dot{x}_i(t) = \frac{p_i(t)}{m_i} \quad (3.9)$$

$$\dot{p}_i(t) = -\frac{\partial V_i^{SCF}(x_i)}{\partial x_i}\bigg|_{x_i(t)} \quad (3.10)$$

$$\dot{A}_i(t) = -\frac{2}{m_i}A_i(t)^2 - \frac{1}{2}\frac{\partial^2 V_i^{SCF}}{\partial x_i^2}\bigg|_{x_i(t)} \quad (3.11)$$

$$\dot{\gamma}_i(t) = \frac{i\hbar A_i(t)}{m_i} + \frac{p_i^2}{2m_i} - V_i^{SCF}(x_i(t)) \quad (3.12)$$

where we have assumed that $H_i(x_i) = -(\hbar^2/2m_i)(\partial^2/\partial x_i^2)$. Thus we obtain classical equations of motion describing the motion of the center of the wavepacket together with an equation for the change of the width $A_i(t)$ and the phase $\gamma_i(t)$. The parameters are, however, not independent, since normalization $\langle\phi_i|\phi_i\rangle = 1$ gives the following constraint:

$$\text{Im } \gamma_i(t) = \frac{\hbar}{4}\ln\left(\frac{\pi\hbar}{2\,\text{Im }A_i(t)}\right) \quad (3.13)$$

The width of the wavepacket is obtained as

$$\Delta x = \frac{1}{2}\sqrt{\frac{\hbar}{\text{Im }A(t)}} \quad (3.14)$$

The momentum space distribution defined by $\int dx \exp(ikx)\phi(x,t)$ is also Gaussian, i.e.,

$$|\phi^k(t)|^2 = \sqrt{\frac{\hbar \operatorname{Im} A(t)}{2\pi|A(t)|^2}} \exp\left(-\frac{\hbar \operatorname{Im} A(t)}{2|A(t)|^2} (k - p(t)/\hbar)^2\right) \tag{3.15}$$

with a width in momentum space

$$\Delta p = \hbar \Delta k = |A(t)| \sqrt{\frac{\hbar}{\operatorname{Im} A(t)}} \tag{3.16}$$

where k is the wavenumber. Note that the GWP obeys Heisenberg's uncertainty principle since

$$\Delta p \Delta x = \frac{\hbar}{2} \frac{|A(t)|}{\operatorname{Im} A(t)} \geq \frac{\hbar}{2} \tag{3.17}$$

Only in the case where $\operatorname{Re} A = 0$ do we get a minimum uncertainty wavepacket, indicated by the equality sign in the above equation.

If the potential is quadratic, the GWP is an exact solution to the TDSE. For example considering a parabolic barrier we have

$$V(x) = V_0 - \tfrac{1}{2}m\omega^2 x^2 \tag{3.18}$$

for $|x| < x_0$; otherwise $V(x) = 0$. The force constant has been expressed in terms of mass and frequency as for the harmonic oscillator. The solution for $|x| < x_0$ is

$$x(t) = x(0) \cosh(\omega t) + \frac{p(0)}{m\omega} \sinh(\omega t) \tag{3.19}$$

$$p(t) = p(0) \cosh(\omega t) + m\omega x(0) \sinh(\omega t) \tag{3.20}$$

$$A(t) = \frac{m\omega}{2} \frac{2A(0) + m\omega \tanh(\omega t)}{m\omega + 2A(0) \tanh(\omega t)} \tag{3.21}$$

The solution for the case of the harmonic oscillator [15] can be obtained in a similar fashion. If the potential $V(x)$ is zero the solution is

$$x(t) = x(t_0) + p(t_0)(t - t_0) \tag{3.22}$$

$$p(t) = p(t_0) \tag{3.23}$$

and

$$A(t) = \frac{A_0}{1 + \dfrac{2A_0}{m} (t - t_0)} \tag{3.24}$$

where $A_0 = A(t_0)$. The expression for $A(t)$ shows that the wavepacket spreads with time since $A(t) \to 0$ and therefore $\Delta x \to \infty$ [see Eq. (3.14)].

Even without the Gaussian wavepacket approximation the TDSCF method introduces a drastic reduction in the complexity since it reduces the problem to a mode-separable one. The TDSCF conserves the average energy, i.e.,

$$\frac{\partial}{\partial t} \langle \Psi | H | \Psi \rangle = 0 \tag{3.25}$$

Although the dynamics of each mode can be solved separately, the method does allow for some coupling (energy exchange) between the modes. The coupling is of an average type; therefore the separation of the modes in this manner can be expected to hold if the coupling is weak or extremely localized in time. Thus the correlation between various modes is not described correctly in the time-dependent SCF method. Note, however, that by introducing a GWP assumption in each mode and using the approximation

$$\langle \phi_i(x_i, t) | V(x_i, x_j, t) | \phi_i(x_i, t) \rangle \sim V(\langle x_i \rangle, x_j, t) \tag{3.26}$$

for each mode we obtain the classical equations of motion, since $\langle x_i \rangle = \langle \phi_i | x_i | \phi_i \rangle = x_i(t)$ in the GWP description. Since classical mechanics does include correlation effects between the various modes, we see that in the SCF approximation we treat the correlation classically if yet another approximation, Eq. (3.26), "the central force" approximation [16], is introduced.

EXERCISES

3.1. Show that Eq. (3.13) gives normalization of the Gaussian wavepacket at all times.

3.2. Obtain the equations of motion for a Gaussian wavepacket in a harmonic potential. What is the condition for a minimum uncertainty wavepacket?

3.2   THE CLASSICAL PATH EQUATIONS

The so-called classical path equations [17] can be obtained from a product-type wave function by letting the trial function in the "classical" degree of freedom be a Gaussian wavepacket. For a system with two degrees of freedom, r and R, with the Hamiltonian

$$H = \hat{H}_0(r, \hat{p}_r) - \frac{\hbar^2}{2\mu} \frac{\partial^2}{\partial R^2} + V(r, R) \tag{3.27}$$

we would use

$$\psi(R, r, t) = \phi(r, t) \exp\left\{ \frac{i}{\hbar} \left[\gamma(t) + P_R(t)(R - R(t)) + A(t)(R - R(t))^2 \right] \right\} \tag{3.28}$$

If we insert this trial function in the TDSE and equate equal powers of $(R - R(t))^k$ with $k = 0$, 1, and 2, we obtain the following equations

$$\dot{R}(t) = \frac{P_R(t)}{\mu} \tag{3.29}$$

$$\dot{P}_R = -\frac{\partial}{\partial R} \langle \phi | V(r, R) | \phi \rangle |_{R = R(t)} \tag{3.30}$$

$$\dot{A} = -\frac{2}{\mu} A^2 - \frac{1}{2} \frac{\partial^2}{\partial R^2} \langle \phi | V(r, R) | \phi \rangle |_{R = R(t)} \tag{3.31}$$

$$\dot{\gamma} = \frac{i\hbar A}{\mu} + \frac{P_R^2}{\mu} \tag{3.32}$$

where the brackets indicate integration over r. For the quantum part of the system we obtain the following TDSE

$$i\hbar \frac{\partial \phi}{\partial t} = [\hat{H}_0 + V(r, R(t)) + F(t)]. \tag{3.33}$$

where $F(t)$ is a phase factor

$$F(t) = \dot{\gamma} - \frac{P_R(t)^2}{2\mu} - \frac{i\hbar A(t)}{\mu} = P_R(t)^2/2\mu. \tag{3.33a}$$

Thus we see that the classical path equations used in the previous section involve the simultaneous integration of the "classical" equations of motion Eqs. (3.29) and (3.30) and the time-dependent Schrödinger equation, Eq. (3.33). The equations are coupled self-consistently through the effective potential.

It is possible to include corrections to the classical path equations by multiplying the above trial function with a correction factor $\sum_{k=0} \varepsilon_k(t)(R - R(t))^k$

so as to be able to fulfill the TDSE to all powers of $(R - R(t))$ [20]. Other corrections to the SCF scheme will be considered below.

3.3 CORRECTIONS TO THE SCF APPROACH

There are several cases, where the SCF approach is known to be inadequate [24]. One such case is the description of tunneling through a barrier, where the tunneling motion is coupled to a heat bath consisting of harmonic oscillators. The following model potential has been used by Makri and Miller [19] and Billing and Jolicard [21]

$$V(s, Q) = -\tfrac{1}{2}a_0 s^2 + \tfrac{1}{4}c_0 s^4 + \tfrac{1}{2}kQ^2 - f(s)Q \qquad (3.34)$$

where s is the tunneling coordinate and Q the oscillator coordinate. The potential in s represents a model for the double-well problem with a barrier at $s = 0$. Introducing a product-type wave function $\Psi(s, Q, t) = \psi_1(s, t)\psi_2(Q, t)$ we get

$$i\hbar\, \frac{\partial \psi_1}{\partial t} = \langle \psi_2 | V(s, Q) | \psi_2 \rangle \psi_1(s, t) - \frac{\hbar^2}{2m} \frac{\partial^2}{\partial s^2}\, \psi_1(s, t) \qquad (3.35)$$

$$i\hbar\, \frac{\partial \psi_2}{\partial t} = -\frac{\hbar^2}{2M} \frac{\partial^2}{\partial Q^2}\, \psi_2(Q, t) + \langle \psi_1 | V(s, Q) | \psi_1 \rangle \psi_2(Q, t) \qquad (3.36)$$

where m and M are the masses of the tunneling particle and the heat bath atoms respectively.

The brackets indicate integration over Q and s respectively. Assuming the potential given above we notice that Eq. (3.36) is the Schrödinger equation for a linearly forced harmonic oscillator and therefore the solution is as given in Chapter 2. The problem with the SCF method in this case, is that the tunneling degree of freedom only feels the average rather than the instantaneous force from the "heat" bath. It has been shown [19, 21] that the single configuration approach gives an adequate description for only a short time period and then breaks down. In order to improve the agreement with exact calculations, one can introduce more configurations [19]. Nevertheless it is also possible to improve the SCF approach by introducing ways of optimizing the parameters of the single configuration wave function [21, 24].

In general, however, the correct expansion of the wave function would allow for mixing of the modes at all times, i.e.,

$$\Psi(x_1, x_2, t) = \sum_{n_1 n_2} c_{n_1 n_2}(t)\phi_{n_1}(x_1)\phi_{n_2}(x_2) \qquad (3.37)$$

where the basis set is stationary and all the time dependence is put into the expansion coefficients. It is, for many-dimensional problems, convenient (in order to reduce the basis set) to let the basis functions themselves evolve in time, according to the dynamics of the system. Thus we would expand the wave function in trial functions $\phi_j(x, t)$

$$\Psi = \sum_{n_1 n_2} c_{n_1 n_2}(t) \phi_{n_1}(x_1, t) \phi_{n_2}(x_2, t) \tag{3.38}$$

Since the time dependence appears both in the expansion coefficients and in the basis functions we can introduce the constraints

$$\langle \phi_{n_i} | \phi_{n_i'} \rangle = \delta_{n_i n_i'} \tag{3.39}$$

$$\langle \phi_{n_i} | \dot{\phi}_{n_i'} \rangle = 0 \tag{3.40}$$

where the integration is over x_i. These constraints simplify the subsequent equations. From the TDSE we obtain the following equations:

$$i\hbar \sum_{n_1 n_2} (\dot{c}_{n_1 n_2} \phi_{n_1} \phi_{n_2} + c_{n_1 n_2} \dot{\phi}_{n_1} \phi_{n_2} + c_{n_1 n_2} \phi_{n_1} \dot{\phi}_{n_2}) = \hat{H} \sum_{n_1 n_2} c_{n_1 n_2} \phi_{n_1} \phi_{n_2} \tag{3.41}$$

By multiplying from the left with $\langle \phi_{m_1} \phi_{m_2} |$ and integrating over the coordinates x_1 and x_2 we obtain a set of equations for the expansion coefficients

$$i\hbar \frac{dc_{m_1 m_2}(t)}{dt} = \sum_{n_1 n_2} \langle \phi_{m_1} \phi_{m_2} | \hat{H} | \phi_{n_1} \phi_{n_2} \rangle c_{n_1 n_2}(t) \tag{3.42}$$

where the above constraints have been used and the brackets denote integration over coordinate space. In order to obtain equations for the evolution of the basis functions, we multiply with $\langle \phi_{j_2} |$ from the left, integrate over x_2, and insert Eq. (3.42) into Eq. (3.41), to obtain

$$i\hbar \sum_{n_1} c_{n_1 j_2}(t) \frac{\partial}{\partial t} \phi_{n_1}(x_1, t) = \sum_{m_1 m_2} c_{m_1 m_2}(t) \left[\hat{H}_{j_2 m_2}(x_1, t) \phi_{m_1}(x_1, t) \right.$$

$$\left. - \sum_{n_1} H_{n_1 j_2; m_1 m_2}(t) \phi_{n_1}(x_1, t) \right] \tag{3.43}$$

Here

$$\hat{H}_{j_2 m_2}(x_1, t) = \langle \phi_{j_2} | \hat{H} | \phi_{m_2} \rangle \tag{3.44}$$

is an operator, since the integration is carried out over x_2 only, whereas $H_{n_1 n_2; m_1 m_2}$ is a matrix element. The Eq. (3.43) can be written in matrix form if we introduce the projection operator

$$P_1 = \sum_{n_1} |\phi_{n_1} \rangle \langle \phi_{n_1}| \tag{3.45}$$

and use the relation

$$\sum_{m_1 m_2} P_1 \hat{H}_{j_2 m_2} c_{m_1 m_2} |\phi_{m_1} \rangle = \sum_{m_1 m_2 n_1} H_{n_1 j_2; m_1 m_2} c_{m_1 m_2} |\phi_{n_1} \rangle \tag{3.46}$$

where we have introduced the ket-notation for simplicity. Thus we have

$$i\hbar \sum_{n_1} c_{n_1 j_2} |\dot{\phi}_{n_1} \rangle = \sum_{m_1 m_2} (1 - P_1) \hat{H}_{j_2 m_2} |\phi_{m_1} \rangle c_{m_1 m_2} \tag{3.47}$$

or in matrix form

$$i\hbar \dot{\phi}_1 = (\mathbf{c}^T)^{-1}(1 - P_1) \hat{\mathbf{H}}_1 \mathbf{c}^T \phi_1 \tag{3.48}$$

which $\mathbf{c}$ is a matrix, and where we have introduced the notation $\hat{\mathbf{H}}_1$ for the operator and ϕ_1 for the vector containing the basis functions ϕ_{n_1}. A similar equation can be obtained for ϕ_2. We note that the number of basis functions can be reduced by letting the basis set be dynamic, i.e., letting it change in time according to the scheme above, however the numerical effort increases in each time step. The methodology can be extended to many dimensions, and the basis functions can be given a discrete representation, on a grid [18] (see Chapter 5). Other methods that reduce the basis set according to the dynamics involved are the RRGM method (see Chapter 11) and the time-dependent wave operator method [52].

3.4   WAVE OPERATOR APPROACH

Methods based on state expansion techniques have poor convergence properties and will, in the end, have to be abandoned when the system becomes large.

Thus alternative methods must try to circumvent this problem. One such method is the Bloch wave operator approach [25] recently formulated for molecular dynamics problems by Jolicard [52]. In this method, the space is divided in two, an active space S_0 of dimension M and the complementary space S_0^+. The projection operators are P_0 which projects onto the active space and Q_0 onto the complementary space, i.e.,

$$P_0 + Q_0 = I \tag{3.49}$$

where I is the identity operator and

$$P_0 P_0 = P_0 \tag{3.50}$$

$$P_0 Q_0 = 0 \tag{3.51}$$

A wave operator $\Omega(t, t_0)$ is now defined as

$$\Omega(t, t_0) = U(t, t_0) P_0 [P_0 U(t, t_0) P_0]^{-1} \tag{3.52}$$

where $U(t, t_0)$ is the evolution operator for the Hamiltonian $H(t, t_0)$, thus the time-dependent Schrödinger equation is

$$i\hbar \frac{\partial}{\partial t} \Psi(t) = H(t, t_0) \Psi(t) \tag{3.53}$$

The operator $\Omega(t, t_0)$ can be expressed as

$$\Omega(t, t_0) = P_0 + X(t, t_0) \tag{3.54}$$

where $X(t, t_0)$ describes the transitions between the active space S_0 and the complementary space. Thus for a system with three states in active space and five in the complementary space, X has the form

$$X = \begin{matrix} S_0 \\ \\ \\ \\ S_0^+ \\ \\ \\ \end{matrix} \begin{bmatrix} 0 & 0 & 0 & 0 & 0 & 0 & 0 & 0 \\ 0 & 0 & 0 & 0 & 0 & 0 & 0 & 0 \\ 0 & 0 & 0 & 0 & 0 & 0 & 0 & 0 \\ x & x & x & 0 & 0 & 0 & 0 & 0 \\ x & x & x & 0 & 0 & 0 & 0 & 0 \\ x & x & x & 0 & 0 & 0 & 0 & 0 \\ x & x & x & 0 & 0 & 0 & 0 & 0 \\ x & x & x & 0 & 0 & 0 & 0 & 0 \end{bmatrix} \tag{3.55}$$

where x indicate nonzero matrix elements. Using that the projection operators are given as:

$$P_0 = \begin{bmatrix} 1 & 0 & 0 & 0 & 0 & 0 & 0 & 0 \\ 0 & 1 & 0 & 0 & 0 & 0 & 0 & 0 \\ 0 & 0 & 1 & 0 & 0 & 0 & 0 & 0 \\ 0 & 0 & 0 & 0 & 0 & 0 & 0 & 0 \\ 0 & 0 & 0 & 0 & 0 & 0 & 0 & 0 \\ 0 & 0 & 0 & 0 & 0 & 0 & 0 & 0 \\ 0 & 0 & 0 & 0 & 0 & 0 & 0 & 0 \\ 0 & 0 & 0 & 0 & 0 & 0 & 0 & 0 \end{bmatrix} \tag{3.56}$$

and $Q_0 = I - P_0$ we see that $Q_0 X = X$ and $X P_0 = X$. Thus we get the reduced wave operator $X = Q_0 \Omega P_0 = Q_0 U P_0 (P_0 U P_0)^{-1}$. Introducing in addition the reduced evolution operator

$$\tilde{U} = P_0 U P_0 \tag{3.57}$$

we obtain from the TDSE for U that

$$i\hbar \frac{\partial}{\partial t} P_0 U P_0 = P_0 H U P_0 = P_0 H \Omega \tilde{U} \tag{3.58}$$

where the above definition of Ω has been used. For the dynamics in the active space we then have

$$i\hbar \frac{\partial \tilde{U}}{\partial t} = H_{\text{eff}} \tilde{U} \tag{3.59}$$

where the dynamics is governed by an effective Hamiltonian

$$H_{\text{eff}} = P_0 H (P_0 + X) \tag{3.60}$$

By inserting in the TDSE for U and using that $U P_0 = (P_0 + X)\tilde{U}$ it is possible (see Exercise **3**) to derive a time-dependent Schrödinger type equation for $X(t, t_0)$, i.e.,

$$i\hbar \frac{\partial X(t, t_0)}{\partial t} = Q_0 (I - X(t, t_0)) H(t, t_0) (I + X(t, t_0)) P_0 \tag{3.61}$$

where I is the identity operator. This equation is nonlinear in X but is, due to the appearance of the projection operators Q_0 and P_0, to be solved in a reduced space. If the coupling between the space S_0 and the complementary space is weak, it is possible to introduce a perturbation treatment. Thus we have

$$X(t, t_0) = X^{(0)}(t, t_0) + X^{(1)}(t, t_0) + \dots \qquad (3.62)$$

where $X^{(0)}$ is obtained from Eq. (3.61) with $X = 0$ inserted on the right-hand side of the equation, etc. The active space should consist of states that are strongly coupled to each other and to the initial state. In many situations, as for example molecular systems in a strong laser field, it turns out that the number of states which are strongly coupled by the field is small compared with the total number of molecular states [27]. The wave operator approach is very efficient in handling such situations.

EXERCISE

3.3. Derive Eq. (3.61).

4

SEMICLASSICAL THEORIES

Given the large masses of nuclei it is natural to assume that nuclear motion can be described using classical trajectories. The solution of the classical equations of motion does not give rise to problems even for many-particle systems. However, we know as well that quantum effects such as tunneling and interference, among others are important in a number of situations. It is therefore natural to attempt to account for these quantum effects at least approximately while retaining the computational advantage of classical propagation schemes. Several such quantum-classical or semiclassical hybrid schemes have been proposed. We have in the previous chapter mentioned the Gaussian wavepacket and the classical path methods. In the present chapter we shall discuss theories which are based on a semiclassical evaluation of the quantum propagators for specific transitions. The Feynman path integral formulation is a natural starting point for the introduction of the semiclassical limit.

4.1   FEYNMAN PATH INTEGRALS

The Feynman path integral formulation is as mentioned one way of formulating quantum mechanics such that the classical limit is immediately visible [28]. Formally the approach involves the introduction of a quantity S which has a definition resembling that of an action integral. If we for the sake of simplicity consider just a one-dimensional system, i.e., a particle with mass m moving along the x-axis under a potential $V(x, t)$, the quantity S is defined by

$$S(x_2 t_2; x_1 t_1) = \int_{t_1}^{t_2} L(\dot{x}, x, t) \, dt \qquad (4.1)$$

where L is the Lagrangian, i.e.,

$$L = \frac{m}{2}\,\dot{x}^2 - V(x,t) \tag{4.2}$$

and $\dot{x} = dx/dt$. The connection to the quantum mechanical formulation is made through the equation

$$\psi(x_2,t_2) = \int_{-\infty}^{\infty} K(x_2,t_2;x_1,t_1)\psi(x_1,t_1)\,dx_1 \tag{4.3}$$

where $K(x_2,t_2;x_1,t_1)$ is the propagator, i.e., the amplitude for a particle going from the position x_1 at time t_1 to x_2 at t_2. By assuming or integrating over all such "transition" amplitudes multiplied with the amplitude for a given initial position—the wave function—we get the amplitude for a final position, namely $\psi(x_2,t_2)$. We now postulate that the propagator can be written as an integral over all "paths" leading from (x_1,t_1) to (x_2,t_2) each weighted by a phase factor

$$\exp\left[\frac{i}{\hbar}\,S(x_2,t_2;x_1,t_1)\right] \tag{4.4}$$

Thus one postulates that

$$\psi(x_2,t_2) = \frac{1}{N}\int_{-\infty}^{\infty} \exp\left(\frac{i}{\hbar}\,S(x_2,t_2;x_1,t_1)\right)\psi(x_1,t_1)\,dx_1 \tag{4.5}$$

where N is a normalization factor that $t_2 = t_1 + \Delta t$. The justification of the expression and the normalization factor can be found by considering a small time increment $t_2 = t_1 + \Delta t$ and showing that the above expression is (in this limit) the time-dependent Schrödinger equation. Thus the left-hand side is expanded as

$$\psi(x,t) + \Delta t\,\frac{\partial \psi(x,t)}{\partial t} \tag{4.6}$$

where x_2 has been replaced by x, and t_1 by t. We note that the action given by Eq. (4.1) can be approximated as

$$\frac{m}{2}\,\frac{(x-x_1)^2}{\Delta t} - \Delta t\,V(\bar{x},t) \tag{4.7}$$

where $\bar{x}$ is an average distance $\bar{x} = \frac{1}{2}(x+x_1)$. Introducing the variable $\eta = x_1 - x$, and expanding in η and Δt, we get

$$\psi(x,t) + \Delta t \frac{\partial \psi}{\partial t} = \int_{-\infty}^{\infty} d\eta \frac{1}{N} \exp(im\eta^2/2\hbar\Delta t)\left(1 - \frac{i\Delta t}{\hbar} V(x,t)\right)$$
$$\cdot \left(\psi(x,t) + \eta \frac{\partial \psi}{\partial x} + \frac{1}{2}\eta^2 \frac{\partial^2 \psi}{\partial x^2}\right) \tag{4.8}$$

Thus the normalization constant is found by requiring that

$$\frac{1}{N} \int_{-\infty}^{\infty} d\eta \exp(im\eta^2/2\hbar\Delta t) = 1 \tag{4.9}$$

which gives

$$N = \sqrt{\frac{2\pi\hbar\Delta t}{im}} \tag{4.10}$$

By equating terms of order Δt and performing the integrals over the variable η we obtain

$$i\hbar \frac{\partial \psi}{\partial t} = -\frac{\hbar^2}{2m} \frac{\partial^2 \psi}{\partial x^2} + V(x,t)\psi \tag{4.11}$$

i.e., the time-dependent Schrödinger equation. This result can now be generalized to a finite time interval—simply by repetition—such that the propagator becomes

$$K(x_2,t_2;x_1,t_1) = \lim_{\Delta t \to 0} \frac{1}{N^{(n-1)}} \int\int \cdots \int$$
$$\cdot \exp\left(\frac{i}{\hbar} S(x_2t_2;x_1t_1)\right) dy_1\, dy_2 \ldots dy_{n-1} \tag{4.12}$$

The integration (see Fig. 4.1) is carried out over the $n-1$ time segments between the end points x_1 and x_2, i.e., y_i is integrated from $-\infty$ to $+\infty$ for $i = 1,\ldots,$ $n-1$. In some cases this integration can be carried out analytically—namely if the force is at most linear in the coordinate x, and in the general case for Hamiltonians that can be written in a quadratic form. Thus for a free particle the action can be written as

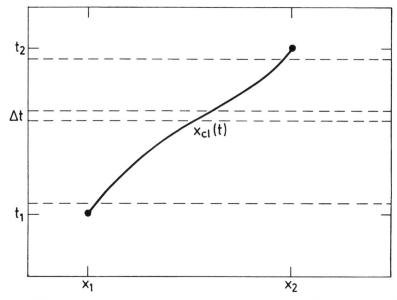

Fig. 4.1. The Feynman path integral can be calculated by dividing the time axis in small intervals Δt and integrating x from $-\infty$ to $+\infty$.

$$S(x_2, t_2; x_1, t_1) = \frac{im}{2\Delta t} \left[\sum_{i=2}^{n-1} (y_i - y_{i-1})^2 + (y_1 - x_1)^2 + (x_2 - y_{n-1})^2 \right] \quad (4.13)$$

Integration over the y_i variables results in the propagator for the free particle

$$K(x_2, t_2; x_1, t_1) = \sqrt{\frac{m}{2\pi i\hbar(t_2 - t_1)}} \, \exp\left(\frac{im(x_2 - x_1)^2}{2\hbar(t_2 - t_1)} \right) \quad (4.14)$$

and for the harmonic oscillator

$$K(x_2, t_2; x_1, t_1) = \sqrt{\frac{m\omega}{2\pi i\hbar \, \sin(\omega T)}} \, \exp\left(\frac{im\,\omega}{2\hbar \, \sin(\omega T)} \right.$$
$$\left. \cdot \left[(x_2^2 + x_1^2) \cos(\omega T) - 2x_1 x_2 \right] \right) \quad (4.15)$$

where $T = t_2 - t_1$. But in general it is not possible to evaluate the propagator. We must therefore resort to considering the evaluation in the short time limit.

4.1.1 The Short-time Propagator

In order to apply the Feynman path concept in actual numerical calculations it is convenient to introduce the short time propagator. Various approximations in the limit of a small Δt have been used. Among those are the Trotter formula [29]

$$K(x_2, t_1 + \Delta t; x_1, t) = \frac{1}{N} \exp\left[\frac{im(x_2 - x_1)^2}{2\hbar\Delta t} - \frac{i\Delta t}{\hbar} \frac{1}{2} (V(x_2) + V(x_1)) \right]$$

(4.16)

However, one can by for example considering the short time propagator for the harmonic oscillator (HO) see that for this particular case one gets (to order Δt) an exponent

$$\frac{im(x_2 - x_1)^2}{2\hbar\Delta t} - \frac{1}{6} \frac{i\Delta t m \omega^2}{\hbar} (x_2^2 + x_1^2 + x_1 x_2)$$

(4.17)

Comparing with the Trotter formula we see that the first term is identical but that the latter is replaced by

$$-\frac{1}{4} \frac{i\Delta t m \omega^2}{\hbar} (x_2^2 + x_1^2)$$

(4.18)

noting that for the harmonic oscillator $V(x) = \frac{1}{2} m \omega^2 x^2$. It has recently been demonstrated by Makri and Miller that the proper short time propagator for small but finite Δt is

$$K(x_2, t_1 + \Delta t; x_1, t) = \frac{1}{N} \exp\left(\frac{im(x_2 - x_1)^2}{2\hbar\Delta t} - \frac{i\Delta t \langle V \rangle}{\hbar} \right)$$

(4.19)

where

$$\langle V \rangle = \frac{1}{\Delta x} \int_{x_1}^{x_2} dx \ V(x)$$

(4.20)

For higher order terms of order Δt^3 the reader is referred to ref. [34]. It can easily be demonstrated that expression in Eq. (4.19) for the harmonic oscillator gives the correct Δt term.

4.1.2   The Classical Limit

The action S appearing in the Feynman path integral formulation can be used to define a trajectory $\bar{x}(t)$ by introducing a least action principle, i.e., the trajectory is defined by

$$\delta S = S[\bar{x} + \delta x] - S[\bar{x}] = 0 \qquad (4.21)$$

By introducing the definition of the action integral it is possible to show that the least action principle leads to the Lagrangian equations of motion:

$$\frac{d}{dt}\frac{\partial L}{\partial \dot{x}} - \frac{\partial L}{\partial x} = 0 \qquad (4.22)$$

In the classical limit one expects that most of the contribution to the path integral comes from the region close to the classical path. In the limit $\hbar \rightarrow 0$, contributions from paths farther away from the one defined by $\delta S = 0$ are small due to the rapid oscillations of the integrand. It is then convenient to expand the action around this path, i.e.,

$$S[x(t)] = S[\bar{x}(t) + y(t)] = \int_{t_1}^{t_2} dt\, \frac{m}{2}\, (\dot{\bar{x}} + \dot{y})^2 - V[\bar{x}(t) + y(t)] \qquad (4.23)$$

Since the classical trajectory is defined by requiring that the linear term in the displacement from the path vanishes, this action integral can be written as

$$S[x(t)] = S_{cl}[\bar{x}(t)] + \int_{t_1}^{t_2} dt \left(\frac{m}{2}\, \dot{y}^2 - \frac{1}{2}\left.\frac{\partial^2 V}{\partial x^2}\right|_{\bar{x}(t)} y^2 \right) \qquad (4.24)$$

The propagator can in this approximation be expressed as

$$K(x_2, t_2; x_1, t_1) = \exp\left(\frac{i}{\hbar}\, S_{cl}[\bar{x}(t)] \right) F(t_2, t_1) \qquad (4.25)$$

where we have made use of the assumption that the classical trajectory starts in x_1 and ends in x_2. Hence the integration over the deviation y must be carried out subject to the condition that $y(t_1) = y(t_2) = 0$. Then the path integral over y can only depend on time. Thus in this "semiclassical" limit the propagator consists of the classical action times a time-dependent normalization function $F(t_2, t_1)$. One can furthermore demonstrate that the function $F(t_2, t_1)$ can be obtained from the classical action as [30]

$$F(t_2,t_1) = \left(\frac{1}{2\pi i\hbar} \left| \frac{\partial^2 S_{cl}}{\partial x_1 \, \partial x_2} \right| \right)^{1/2} \tag{4.26}$$

Introducing the Hamiltonian $H(x,p_x) = (m/2)\dot{x}^2 + V(x)$ we can rewrite the classical action as

$$S_{cl}(x_2,x_1,\Delta t) = \int_{x_1}^{x_2} dx \, p_x - H(x_1,p_1)\Delta t \tag{4.27}$$

where $\Delta t = t_2 - t_1$. The semiclassical propagator is exact if the action is a quadratic function of the coordinates x_1 and x_2. In general, it will be a locally correct or a globally approximate propagator.

EXERCISES

4.1. Derive Eq. (4.26) by using that $K(x_2,t_2;x_1,t_1)$ should be unitary, i.e. that

$$\int dx_1 \, K^*(x_2,t_2;x_1,t_1)K(x,t_2;x_1,t_1) = \delta(x - x_2) \tag{4.28}$$

and

$$\int_{-\infty}^{\infty} dx \, \exp(ixy) = 2\pi\delta(y) \tag{4.29}$$

4.2. Use Eq. (4.26) to obtain the normalization factor for a free particle and the harmonic oscillator.

4.2 SEMICLASSICAL FEYNMAN PATH THEORY

As we have seen the path integral formulation is easy to use if deviations up to second order from the classical path are included or if it is used in the short time limit. It is therefore possible to suggest an approach where part of the system, e.g., the internal degrees of freedom, are treated quantum mechanically, whereas the translational degree of freedom is treated using the path integral formulation. Such a hybrid theory was suggested by Pechukas in 1969 [31]. Consider a system consisting of the internal (vibrational) coordinate r and the external translational coordinate R. Thus for this system the Lagrangian is

$$L(r,R) = \frac{\mu}{2}\,\dot{R}^2 + \frac{m}{2}\,\dot{r}^2 - v(r) - V(r,R) \tag{4.30}$$

where μ is the reduced mass for the relative motion, and m the reduced mass of the diatomic molecular, $v(r)$ the intramolecular and $V(r,R)$ the intermolecular potentials respectively. The propagator $K(R_2,r_2,t_2;R_1,r_1,t_1)$ satisfies the Schrödinger equation, i.e.,

$$i\hbar\,\frac{\partial K(R_2,r_2,t_2;R_1,r_1,t_1)}{\partial t_2} = \left[-\frac{\hbar^2}{2\mu}\,\frac{\partial^2}{\partial R_2^2} - \frac{\hbar^2}{2m}\,\frac{\partial^2}{\partial r_2^2} + V(R_2,r_2) + v(r_2)\right]$$

$$\cdot K(R_2,r_2,t_2;R_1,r_1,t_1) \tag{4.31}$$

and the boundary condition

$$K(R_2,r_2,t_2;R_1,r_1,t_1) \rightarrow \delta(R_2 - R_1)\delta(r_2 - r_1) \text{ for } t_2 \rightarrow t_1 \tag{4.32}$$

Introducing the eigenstates to the Hamiltonian

$$H_0 = -\frac{\hbar^2}{2m}\,\frac{\partial^2}{\partial r^2} + v(r)$$

i.e.,

$$H_0(r)\phi_n(r) = E_n\phi_n(r) \tag{4.33}$$

we can introduce the amplitude K_{mn} as

$$K_{mn}(R_2,t_2;R_1,t_1) = \int dr \int dr'\,\phi_m(r')^* K(R_2,r',t_2;R_1,r,t_1)\phi_n(r) \tag{4.34}$$

K_{mn} is the amplitude for the system going from state n and separation R_1 to state m and separation R_2 during the time interval $t_2 - t_1$. The Feynman representation for this reduced propagator is

$$K_{mn}(R_2,t_2;R_1,t_1) = \int DR(t)\exp\left(\frac{i}{\hbar}\,S([R(t)])\right) T_{mn}([R(t)]) \tag{4.35}$$

where $DR(t)$ denotes a path integral, $S([R(t)])$ is the action

$$S([R(t)]) = \int_{t_1}^{t_2} dt\,\frac{\mu\dot{R}(t)^2}{2} \tag{4.36}$$

and T_{mn} the amplitude for a transition from quantum state n to m induced by the trajectory $R(t)$. This transition amplitude can be found using the traditional solution of the TDSE (see Section 2.1). In order to investigate the influence of the quantum transition on the classical trajectory, it is interesting to determine the semiclassical trajectory as the path $R(t)$, which makes the phase of Eq. (4.35) stationary, i.e.,

$$\delta[S([R(t)]) + \hbar \, \text{Im} \, \ln T_{mn}([R(t)])] = 0 \qquad (4.37)$$

This equation defines the path $R(t)$, and the equation of motion which has to be solved in order to obtain the path is [31]

$$\mu \frac{d^2 R(t)}{dt^2} = \text{Re} \left[\frac{\left\langle \psi_m(t_2, t) \left| -\frac{\partial V}{\partial R} \right| \psi_n(t, t_1) \right\rangle}{\langle \psi_m(t_2, t) | \psi_n(t, t_1) \rangle} \right] \qquad (4.38)$$

i.e., the trajectory at time t is obtained by solving the usual equation of motion with an average force—averaged over a wave function ϕ_n propagated forward in time from t_1 to t, and a wave function ϕ_m propagated backwards in time from t_2 to t. Since the evolution operator that governs this propagation does depend (through $V(r, R)$) on the trajectory in the time intervals above, we see that the problem is noncausal, i.e., has to be solved iteratively [32]. It can be shown [31] that the trajectory conserves energy in the "quantum mechanical" sense, i.e., that we for $t_1 \rightarrow -\infty$ and $t_2 \rightarrow \infty$ have

$$\mu \frac{\dot{R}(t_1)^2}{2} + E_n = \mu \frac{\dot{R}(t_2)^2}{2} + E_m \qquad (4.39)$$

Thus each quantum mechanical transition has its own trajectory. The theory is most easily used in the short time limit where there is only one path. Such a method has recently been used to determine classical paths under nonadiabatic electronic excitations [33].

4.3   CLASSICAL S-MATRIX THEORY

The classical S-matrix theory was formulated in the beginning of the seventies primarily by Miller and Marcus [35]. It can be derived from the semiclassical limit of the Feynman propagator.

We have seen above that the propagator can be simplified in the semiclassical limit. Using this limit of the propagator, the quantum corrections to the classical trajectories are immediately visible. Let us then use for the final momentum the expression

$$\frac{\partial S_{cl}}{\partial x_2} = p_2 \tag{4.40}$$

and extend the discussion to two degrees of freedom, i.e., that of a collinear collision between an atom (A) and an oscillator (BC). For such a system the Hamiltonian is

$$H = \frac{P_R^2}{2\mu} + V(R, r) + \frac{p_r^2}{2m} + v(r) \tag{4.41}$$

where μ is the reduced mass $\mu = m_A(m_B+m_C)/(m_A+m_B+m_C)$; $m = m_B m_C/(m_B + m_C)$; $V(R, r)$ is the intermolecular potential; and $v(r)$ the intramolecular potential depending on the $B - C$ bond length. It is then convenient to introduce the action-angle variables for the oscillator (see Exercise **3**). This allows us to rewrite the Hamiltonian as

$$H = \frac{P_R^2}{2\mu} + V(R, \phi, t) + n\omega \tag{4.42}$$

where n is the classical vibrational action and ϕ the corresponding angle. Thus by analogy to Eq. (4.27)

$$S_{cl} = \int dR \; P_R + \int n \; d\phi - H(t_2 - t_1) \tag{4.43}$$

and we also have

$$\frac{\partial S_{cl}}{\partial \phi_2} = n_2 \tag{4.44}$$

Introducing the action operator $\hat{n}$ such that $[\hat{n}, \phi] = -i\hbar$ we see that $\hat{n} = -i\hbar(\partial/\partial\phi)$ and that the wave function can be introduced as $\psi(\phi) = \exp(in\phi/\hbar)$ [36] since

$$\hat{n}\psi(\phi) = n\psi(\phi) \tag{4.45}$$

Thus we can express the initial wavefunction as

$$\Psi_{n_1}(R_1, \phi_1) = \sqrt{\frac{1}{2\pi v_1}} \; \exp(in_1^0\phi_1 + iP_1^0 R_1/\hbar) \tag{4.46}$$

where $v_1 = P_1^0/\mu$ is the initial velocity, $R_1 = R(t_1)$ and

$$E = \hbar(n_1^0 + 1/2) + \frac{(P_1^0)^2}{2\mu} \tag{4.47}$$

is the total energy expressed in terms of the initial vibrational quantum number n_1^0 and the kinetic energy of the relative motion.

The final wave function can be expanded in the following manner

$$\Psi(R_2, \phi_2) = \frac{1}{\sqrt{2\pi v_2}} \sum_{n_2^0} S_{12} \exp(in_2^0\phi_2 + iP_2^0R_2/\hbar) \tag{4.48}$$

where we have written the S-matrix elements $S_{n_1^0;n_2^0}$ as S_{12}. The final wave function can also be written as

$$\Psi(R_2, \phi_2, t_2) = \int dR_1 \, d\phi_1 \, K(R_2, \phi_2, t_2; R_1, \phi_1, t_1) \Psi_{n_1}(R_1, \phi_1, t_1) \tag{4.49}$$

We then introduce the semiclassical propagator from Eq. (4.25) and rewrite the classical action as

$$S_{cl} = R_2P_R(t_2) - R_1P_R(t_1) - \int R \, dP_R + \int n \, d\phi - E(t_2 - t_1) \tag{4.50}$$

Here we have introduced that for a conservative system $H = E$. Using the above equations along with Eq. (4.49) we obtain the following S-matrix elements

$$S_{12} = \delta(P_R(t_1) - P_1^0)\delta(P_R(t_2) - P_2^0) \int d\phi_2 \int d\phi_1 \left(\frac{1}{2\pi i\hbar} \frac{\partial n_2}{\partial \phi_1} \right)^{1/2} \frac{1}{2\pi}$$

$$\cdot \exp(in_1^0\phi_1 - in_2^0\phi_2 + iS_{cl}^*/\hbar) \tag{4.51}$$

where we have used Eq. (4.44) and that the translational wave functions are normalized. We have furthermore introduced S_{cl}^*

$$S_{cl}^* = -E(t_2 - t_1) - \int R \, dP_R + n_2\phi_2 - n_1\phi_1 - \int \phi \, dn \tag{4.52}$$

Evaluation of this integral requires knowledge about all trajectories selected from the initial phase space ϕ_1. The final action $n(t_2) = n_2$, as well as the classical action S_{cl}^*, will be functions of the initial phase. The double integral may be reduced to a single one by evaluating the ϕ_2 integral using the stationary phase approach, i.e., expanding the phase Φ

$$\Phi \sim \Phi^0 + \left.\frac{\partial \Phi}{\partial \phi_2}\right|_0 (\phi_2 - \phi_2^0) + \frac{1}{2}\left.\frac{\partial^2 \Phi}{\partial \phi_2^2}\right|_0 (\phi_2 - \phi_2^0)^2 \qquad (4.53)$$

Using that $\partial^2 \Phi/\partial \phi_2^2 = \partial n_2/\partial \phi_2$ we get

$$S_{12} = \delta(P_R(t_1) - P_1^0)\delta(P_R(t_2) - P_2^0)\delta(n_2 - \hbar n_2^0)\frac{1}{2\pi}\int_0^{2\pi} d\phi_1 \left.\left(\frac{\partial \phi_2}{\partial \phi_1}\right)^{1/2}\right|_{n_1^0}$$

$$\cdot \exp\left(\frac{i}{\hbar}\left[(n_1^0\hbar - n_1)\phi_1 - \int_{P_{R1}}^{P_{R2}} R\, dP_R - \int_{n_1}^{n_2} \phi\, dn\right]\right) \qquad (4.54)$$

where the δ function arises from the stationary phase condition $n_2 = \hbar n_2^0$. The initial conditions are

$$n_1(t_1) = \hbar(n_1^0 + 1/2) \qquad (4.55)$$

$$\phi_1 \text{ random in } [0; 2\pi] \qquad (4.56)$$

$$R_1 = R(t_0) \text{ large} \qquad (4.57)$$

$$P_R(t_1) = -\sqrt{2\mu(E - n_1\omega)} \qquad (4.58)$$

where n_1^0 is the initial vibrational quantum number and where the usual Langer correction $(+1/2)$ has been introduced. In classical S-matrix theory one searches for the trajectories selected from the phase-space ϕ_1 which connects the initial n_1^0 and final quantum number n_2^0. The solution of this double-ended boundary value problem is known as root-searching. If this problem has no solution the transition is said to be classically forbidden, i.e., no classical trajectory connects the two quantum states. However, it is possible to extend the theory to complex space and search for solutions for complex values of the angle ϕ_1. The evaluation of the above integral can be performed if the "roots" in ϕ_1 space are well separated, using the stationary phase method for each root. Using that

$$\int_{-\infty}^{\infty} \exp(\pm iax^2) = \sqrt{\frac{\pi}{a}}\ \exp(\mp i\pi/4) \qquad (4.59)$$

i.e., with two roots we obtain

$$S_{12} \sim \sqrt{P_a} \, \exp(i\Phi_1^a + i\pi/4) + \sqrt{P_b} \, \exp(i\Phi_1^b - i\pi/4) \qquad (4.60)$$

where the phase Φ_1 is evaluated at the two roots. Thus the expression includes interference effects between roots a and b. The probabilities P_a and P_b are obtained as

$$P_a = -\frac{1}{2\pi} \left(\frac{\partial \phi_2}{\partial \phi_1} \right)_{n_1^0} \bigg/ \left(\frac{\partial^2 \Phi_1}{\partial \phi_1^2} \right) \qquad (4.61)$$

with a similar expression for P_b. Using that $i\Phi_1$ is the phase of the integrand in Eq. (4.54), we see that

$$\partial \Phi_1 / \partial \phi_1 = (n_1^0 - n_1/\hbar) - \frac{1}{\hbar} \frac{\partial n_2}{\partial \phi_1} \phi_2 \qquad (4.62)$$

and

$$\frac{\partial^2 \Phi_1}{\partial \phi_1^2} \sim -\frac{1}{\hbar} \frac{\partial n_2}{\partial \phi_1} \frac{\partial \phi_2}{\partial \phi_1} \qquad (4.63)$$

where we have used that the final angle ϕ_2 depends on the initial angle ϕ_1. The roots are defined by the value of ϕ_1 which gives $n_2(\phi_1) = n_2^0$ (see Fig. 4.2). Using Eq. (4.63) we get

$$P_a = \frac{\hbar}{2\pi} \frac{\partial \phi_1}{\partial n_2} \qquad (4.64)$$

Thus the classical probability P_a is given by the slope of the initial phase angle with the final action. Introducing $\Delta n_2 = \hbar$ ("boxing") we see that P_a is a measure of the phase space that gives trajectories belonging to a range around the final state n_2^0. If the two roots are not separated, special techniques have to be used in order to evaluate the integral (see for example [36]).

There are two problems connected to the formulation of the theory as given in this section. They are the doubled-ended boundary problem (the root-searching) and the caustics (the zeroth of the normalizing factor 4.26). In multidimensional problems this factor becomes a determinant [37]. The number ν of the zeroth of the determinant gives rise to a factor $\exp(-i\nu\pi/2)$ [38]. In order to deal with such problems it is possible to reformulate the theory so as to obtain an initial value representation [39]. In the initial value representation, the evaluation of the propagator involves an integration over the initial momentum. However, problems with strong oscillations of the integrand as a function of the initial variable can still appear and make a straightforward evaluation impos-

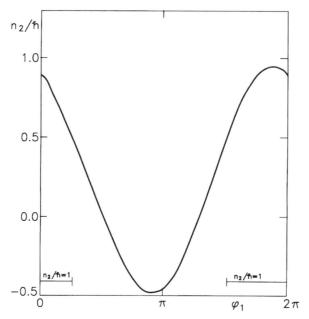

Fig. 4.2. The classical vibrational action n_2 as a function of the initial phase angle ϕ_1. For a given final quantum number n_2^0 we have two, one, or zero roots.

sible. This problem can be circumvented by introducing some smoothing before the integration is carried out. Such smoothing can be introduced in the form of a Filinov-transformation [40].

EXERCISES

4.3. By means of the Bohr-Sommerfeld relation for the action

$$n = \frac{1}{2\pi} \oint p \, dr \tag{4.65}$$

and the classical generator $F_2(r, n)$ where

$$\phi = \frac{\partial F_2}{\partial n} \tag{4.66}$$

$$p_r = \frac{\partial F_2}{\partial r} \tag{4.67}$$

derive an expression for the classical Hamiltonian in terms of action-angle variables for the harmonic oscillator.

4.4   RESOLUTION IN EIGENSTATES AND STATISTICAL MECHANICS

Suppose that the system under consideration can be described by a number of stationary states $\phi_n(x)$, which form a complete orthonormal set, then the wave function $\psi(x, t)$ can be expanded

$$\psi(x, t) = \sum_n c_n \exp(-iE_n t/\hbar)\phi_n(x) \tag{4.68}$$

Inserting this expression in Eq. (4.3) we get

$$\psi(x_2, t_2) = \sum_n \phi_n(x_2) \exp\left(-\frac{i}{\hbar}\, t_2 E_n\right) c_n$$

$$= \int dx_1\, K(x_2, t_2; x_1, t_1) \sum_n c_m \phi_m(x_1) \exp\left(-\frac{i}{\hbar}\, t_1 E_m\right) \tag{4.69}$$

From this equation we immediately see that the propagator $K(x_2, t_2; x_1, t_1)$ can be resolved in eigenstates

$$K(x_2, t_2; x_1, t_1) = \sum_n \phi_n(x_2)\phi_n^*(x_1) \exp\left[-\frac{i}{\hbar}\, E_n(t_2 - t_1)\right] \tag{4.70}$$

where $t_2 > t_1$. This result can now be used in an interesting manner—namely to give some relation to statistical mechanics. Let us consider a system connected to a larger "heat bath" maintained at a given temperature T. We can then assume that the quantum states are distributed according to the Boltzmann law and hence that the probability of finding a given state n is proportional to $\exp(-\beta E_n)$, where $\beta = 1/kT$. The probability of the system having a given value of the coordinate x is

$$P(x, T) = \frac{1}{Q(T)} \sum_n \phi_n^*(x)\phi_n(x) \exp(-\beta E_n) \tag{4.71}$$

where $Q(T)$ is the partition function

$$Q(T) = \int dx\, \rho(x, x) \tag{4.72}$$

and

$$\rho(x',x) = \sum_n \phi_n^*(x')\phi_n(x)\exp(-\beta E_n) \qquad (4.73)$$

We note that this density matrix $\rho(x',x)$ resembles the expression for the propagator above and is in fact identical if we replace $t_2 - t_1$ with $-i\beta\hbar$. Thus it is possible to evaluate the density matrix as a path integral by introducing complex time steps. For the harmonic oscillator we would get

$$\rho(x',x) = \sqrt{\frac{m\omega}{2\pi\hbar\,\sinh(\omega\beta\hbar)}}$$

$$\cdot \exp\left\{-\frac{m\omega}{2\hbar\,\sinh(\omega\beta\hbar)}\,[(x^2 + x'^2)\cosh(\omega\beta\hbar) - 2xx']\right\} \qquad (4.74)$$

from which it is easy to show that

$$Q(T) = \int dx\,\rho(x,x) = \frac{1}{2\sinh(0.5\omega\beta\hbar)} \qquad (4.75)$$

which demonstrates that the correct harmonic oscillator partition function is obtained. Considering the short "time" limit of the propagator we get

$$\rho(x_2,x_1) = \sqrt{\frac{m}{2\pi\beta\hbar^2}}\,\exp\left[-\frac{m(x_2 - x_1)^2}{2\hbar\beta\hbar} - \beta\langle V\rangle\right] \qquad (4.76)$$

If the temperature is not too low (β small) then this short "time" propagator expression is a sufficiently accurate approximation. In any case, since the integrand is not highly oscillatory as in the real time case, the "path-integrals" will converge rapidly. Hence, due to the exponential behavior of the propagator, paths that during the short "time" interval deviate from the region around x_1 will not contribute. Thus the partition function can be approximated by

$$Q(T) \sim \sqrt{\frac{mkT}{2\pi\hbar^2}}\int dx_1\,\exp(-\beta V(x_1)) \qquad (4.77)$$

We note that for large temperatures, small complex time steps are needed and that one, in order to evaluate quantum mechanical corrections to the statistical equilibrium functions, needs only the trace of the density matrix, i.e., in the path integral sense all paths have to be closed.

5

WAVEPACKET PROPAGATION

Efficient ways of evaluating the effect of kinetic energy operators on a wave function make it convenient to propagate the wave function in time on a space grid. On the grid, the wave function is represented in a discrete manner and the potential is "diagonal." Thus the potential operation on the wave function at a given grid point is just obtained by a multiplication with the value of the potential at this grid point. Hence the coupling between the wave function at various grid points occurs through the kinetic energy term. Usual ways of evaluating the kinetic energy terms involve either the fast Fourier transform (FFT) or DVR (discrete variable representation) methods. With these methods it is possible to evaluate the Hamiltonian operator working on the wave function at a given time of the propagation. The time propagation itself can also be carried out in different ways. Thus one can for instance use the split operator, the Chebyshev or the Lanczos time propagation method.

5.1 FOURIER TRANSFORM METHOD

The FFT method is based on the Fourier transform of the wave function

$$\psi(x_j) = \sqrt{\frac{1}{N}} \sum_{k=-N/2}^{N/2-1} c_k \exp(-i2\pi kj/N) \qquad (5.1)$$

where $j = 0, 1, \ldots, N-1$, N the number of gridpoints, $x_j = j\Delta x + 0.5\Delta x$ and $\Delta x = (x_{max} - x_{min})/N$. In Fourier space it is easy to evaluate the kinetic energy terms. Thus we have

$$\frac{\partial^n \psi(x)}{\partial x^n} = \sqrt{\frac{1}{N}} \sum_k c_k \left(-\frac{i2\pi k}{N\Delta x} \right)^n \exp(-i2\pi kj/N) \qquad (5.2)$$

We see that the operators $\partial/\partial x$ working on the wave function can be evaluated by introducing the Fourier coefficients

$$d_k = (\omega_k)^n c_k \qquad (5.3)$$

where the frequency ω_k is given by

$$\omega_k = -\frac{i2\pi k}{N\Delta x} \qquad (5.4)$$

and then making the reverse transform FFT^{-1} with these coefficients. The FFT method is termed fast due to the fact that it is an N logN method in the number of operations and thereby in the computing time as well (see ref. [41]). The output from a complex FFT routine gives a certain order of the frequencies, starting with zero and ending with $-1/N\Delta x$, i.e., we have [41]

$$\omega_j = -\frac{i2\pi}{N\Delta x} j \qquad \text{for } j < \frac{N}{2}$$

$$\omega_j = -\frac{i2\pi}{N\Delta x} (j - N) \qquad \text{for } j \geq \frac{N}{2}$$

where the connections between the j and k index is $j = k + N/2$. If specific boundary conditions need to be fulfilled it is sometimes advantageous to use the DVR method, which however is of order N^2 in the number of operations.

5.2 DVR METHOD

In the DVR method the wave function is also represented at a discrete set of points, the grid points x_m. Between the grid points the function can be evaluated using a suitable interpolation, i.e.,

$$f(x) = \sum_{j=1}^{N} \ell_j(x) f(x_j) + E(x) \qquad (5.5)$$

where $E(x)$ is the error. The interpolation polynomial $\ell_j(x)$ could for example be of the Lagrange type, i.e.,

$$\ell_j(x) = \frac{p_N(x)}{(x - x_j)p'_N(x_j)} \tag{5.6}$$

where

$$p_N(x) = \prod_{i=1}^{N} (x - x_i) \tag{5.7}$$

and p'_N the derivative of p_N. We note that the Lagrange polynomial possesses the property

$$\ell_j(x_k) = \delta_{jk} \tag{5.8}$$

Using this representation of the function $f(x)$ allows us to obtain the derivatives

$$\frac{\partial^k f(x)}{\partial x^k} = \sum_{j=1}^{N} \frac{\partial^k}{\partial x^k} \ell_j(x) f(x_j) + \frac{\partial^k E(x)}{\partial x^k} \tag{5.9}$$

Instead of using the Lagrange function as an interpolation function it is more convenient to use a basis set to construct functions with the above property. In this manner primitive basis functions can be chosen in such a way that specific boundary conditions are fulfilled. As an example we take the basis

$$\psi_n(x) = \sqrt{\frac{2}{\pi}} \sin(nx) \tag{5.10}$$

where $0 \le x \le \pi$, i.e., where the basis functions are zero at the endpoints. The grid points are the zeros of $\psi_{N+1}(x)$, i.e.,

$$x_k = \frac{k\pi}{N + 1} \tag{5.11}$$

where $k = 1, \ldots, N$. We note that an arbitrary wave function $f(x,t)$ can be expanded in the basis set

$$f(x,t) = \sum_{n} c_n(t)\ell_n(x) \tag{5.12}$$

where $\ell_n(x)$ is determined by the primitive basis functions as described below. Thus the expansion coefficients can be evaluated as

$$c_n(t) = \omega_n \int_0^\pi dx \; f(x,t)\ell_n(x) \tag{5.13}$$

where ω_n is a normalization constant and where

$$\ell_n(x) = \frac{2}{N} \sum_{k=1}^N \sin(kx) \; \sin(kx_n) \tag{5.14}$$

The summation over k can be performed

$$\ell_n(x) = \frac{1}{N} \left\{ \cos\left[\frac{N+1}{2}(x-x_n)\right] \sin\left[\frac{N}{2}(x-x_n)\right] \operatorname{cosec}\left(\frac{x-x_n}{2}\right) \right.$$
$$\left. - \cos\left[\frac{N+1}{2}(x+x_n)\right] \sin\left[\frac{N}{2}(x+x_n)\right] \operatorname{cosec}\left(\frac{x+x_n}{2}\right) \right\} \tag{5.15}$$

From this expression we can deduce that the basis functions have the following δ function property: $\ell_n(x_k) \sim \delta_{nk}$. For other values of x the function oscillates near zero. The basis functions $\ell_n(x)$ constitute an orthogonal but not necessarily normalized basis set. Thus we can show that

$$\int_0^\pi dx \; \ell_n(x)\ell_m(x) \sim \frac{\pi}{2N} \delta_{nm} \tag{5.16}$$

As the functions $\ell_n(x)$ have the δ function property then the integral from Eq. (5.13) can be evaluated as

$$c_n(t) = f(x_n, t)\omega_n \tag{5.17}$$

and hence the representation of the wave function becomes

$$f(x,t) = \sum_n \omega_n f(x_n, t)\ell_n(x) \tag{5.18}$$

Derivatives of the wave function are now easy to evaluate,

$$\frac{\partial^k f(x,t)}{\partial x^k} = \sum_n \omega_n f(x_n, t) \frac{\partial^k}{\partial x^k} \ell_n(x) \qquad (5.19)$$

where the basis functions $\ell_n(x)$, as illustrated above, are constructed from an orthonormal basis set $\psi_n(x)$, $(n = 1, \ldots, N)$ such that

$$\ell_n(x) = \sum_i \psi_i^*(x_n) \psi_i(x) \qquad (5.20)$$

and hence the derivatives are obtained as

$$\frac{\partial^k}{\partial x^k} \ell_n(x) = \sum_i \psi_i^*(x_n) \frac{\partial^k \psi_i(x)}{\partial x^k} \qquad (5.21)$$

Thus the evaluation of the derivative at each of the N grid points requires N operations and the DVR method is therefore an N^2 method contrary to the FFT, which as mentioned is an N logN method. However, the situation is slightly more complex, since one in the FFT method should include a constant in front of the N logN and the DVR method involves matrix multiplication, which can utilize the fact that the matrix is sparse and obtain the full speed of vector machines. Another advantage of the DVR method is that the grid points do not have to be evenly spaced but can be placed in regions where the wave function is located.

The number of grid points can be reduced by removing points corresponding to the high-energy components by the sequential adiabatic reduction technique developed by Light and coworkers [43]. This removal of the high-energy components of the spectrum is furthermore important for the time propagation, since the step length of the time propagators is roughly proportional to $\hbar/E_{\max}$.

5.3   TIME PROPAGATION

Time propagation also can follow different paths. The simplest scheme is the second order difference (SOD) scheme. Here we use

$$\phi(x_n, t + \Delta t) = \phi(x_n, t - \Delta t) + \frac{2\Delta t}{i\hbar} \hat{H}(x_n, t)\phi(x_n, t) \qquad (5.22)$$

where $\hat{H}$ denotes the Hamiltonian operator for the system. The SOD method is simple to apply, but requires much smaller time step lengths in order to be more stable than the other methods mentioned below. The error in the SOD method is of order Δt^3. A popular and stable method is the split operator method, in which one uses that

$$\phi(x_n, t + \Delta t) = \exp(-i\Delta t\hat{H}/\hbar)\phi(x_n, t)$$
$$= \exp(i\hbar\nabla^2\Delta t/4m) \ \exp(-iV\Delta t/\hbar) \ \exp(i\hbar\nabla^2\Delta t/4m)\phi(x_n, t)$$

(5.23)

The splitting of the kinetic energy operator makes the method a third order method in Δt, and the exponentiation makes the operator unitary. The operation with the exponential operators on the wave function is evaluated using an FFT or a DVR method. In the FFT method one obtains, by using Eq. (5.1),

$$\exp(i\hbar\nabla^2\Delta t/2m)\phi(x_j)$$

$$= \frac{1}{\sqrt{N}} \sum_k c_k \ \exp\left[\frac{i\hbar\Delta t}{2m}\left(-\frac{i2\pi k}{N\Delta x}\right)^2\right] \exp(-i2\pi kj/N)$$ (5.24)

i.e., the result of operating with the exponential operator on the wave function is found by making a Fourier transform to obtain the c_k coefficients and a back transform with

$$d_k = c_k \ \exp\left(-\frac{i\hbar\Delta t}{2m}\frac{4\pi^2 k^2}{N^2\Delta x^2}\right)$$ (5.25)

The split operator method appears to use two FFTs per time step, but considering two consecutive time steps we have

$$\phi(x_n, t + \Delta t) = \exp(i\hbar\nabla^2\Delta t/4m) \ \exp(-iV\Delta t/\hbar)\exp(i\hbar\nabla^2\Delta t/2m)$$
$$\cdot \ \exp(-iV\Delta t/\hbar) \ \exp(i\hbar\nabla^2\Delta t/4m)\phi(x_n, t - \Delta t)$$ (5.26)

etc. Hence, only one FFT evaluation per time step is needed. However, it is also possible to split the potential instead of the kinetic energy term, i.e., to use

$$\exp(-i0.5V\Delta t/\hbar) \ \exp(i\hbar\nabla^2\Delta t/2m) \ \exp(-i0.5V\Delta t/\hbar)$$ (5.27)

instead. Since the potential is diagonal on the grid, this may be convenient if the Hamiltonian is time-dependent. The method is not restricted to the above simple kinetic energy operator, though complicated operators require more splittings. A method that is capable of taking large time steps (for time-independent Hamiltonians) is the Chebyshev method, where one uses the relation [42]

$$\exp(-i\hat{H}\Delta t/\hbar) = 2 \sum_{k=1}^{\infty} i^k J_k(z) \cos(k\phi) + J_0(z) \tag{5.28}$$

where

$$iz \cos \phi = -i\hat{H}\Delta t/\hbar \tag{5.29}$$

and where J_n is a Bessel function. Introducing now $x = \cos \phi$ and the recursion relations

$$i^k T_k = 2xi^k T_{k-1} - i^k T_{k-2} \tag{5.30}$$

where $T_0 = 1$, $T_1 = \cos \phi$ and $T_k = \cos(k\phi)$ are the Chebyshev polynomials, we obtain

$$S_k(y) = 2yS_{k-1}(y) + S_{k-2}(y) \tag{5.31}$$

where $y = ix$ and $S_k = i^k T_k$. Thus we have

$$zy = -i\hat{H}\Delta t/\hbar \tag{5.32}$$

Since x must be smaller than unity ($x = \cos \phi$) the energy (the Hamiltonian) has to be rescaled in order for the method to converge, i.e., one introduces

$$x = -\hat{H}/E_{max} \tag{5.33}$$
$$z = \Delta t E_{max}/\hbar \tag{5.34}$$

and hence we see that

$$\exp(-i\hat{H}\Delta t/\hbar)\phi = \phi + 2\left(\sum_{n=1}^{\infty} J_n(z)\tilde{\phi}_n - \sum_{n=1}^{\infty} J_{2n}(z)\phi \right) \tag{5.35}$$

where $\tilde{\phi}_n$ is obtained from the recursion formula

$$\tilde{\phi}_n = 2y\tilde{\phi}_{n-1} + \tilde{\phi}_{n-2} \tag{5.36}$$

$\tilde{\phi}_n = S_n\phi$ and where we have used

$$J_0(z) = 1 - 2\sum_{k=1} J_{2k}(z) \tag{5.37}$$

Thus the above scheme is started by evaluating $\tilde{\phi}_0 = \phi$ (the initial wave function) and $\tilde{\phi}_1 = -i\hat{H}\phi/E_{max}$, and using the relation $y = -i\hat{H}/E_{max}$.

The Chebyshev method is convenient to use if the Hamiltonian is time-independent [44]. Here the time interval Δt is taken to be large enough to cover the complete dynamics of the system. The convergence criterion is $J_k(z) \ll 0$. The value of k for which the series may be truncated can be estimated from the asymptotic formula

$$J_k(z) \sim \frac{1}{\sqrt{2\pi k}}\left(\frac{ez}{2k}\right)^k \tag{5.38}$$

If on the other hand the Hamiltonian is time-dependent, either because external fields are included or because part of the system is treated within the classical mechanical approximation, it is more convenient to use the Lanczos reduction technique [45]. The Lanczos reduction technique is yet another method for time propagation. In order to demonstrate the method, we write the TDSE in matrix form

$$i\hbar\dot{\phi} = \mathbf{H}\phi \tag{5.39}$$

where ϕ is a vector of length N, containing the value of the wave function at the N grid points, and $\mathbf{H}$ is an $N \times N$ matrix. We now introduce a transformation matrix that transforms the grid representation of the wave function to a tridiagonal form. The transformation matrix is defined by

$$\mathbf{T} = [\phi_0, \phi_1, \ldots, \phi_{L-1}] \tag{5.40}$$

where $L \ll N$ is the number of Lanczos recursions. The elements of $\mathbf{T}$ are called the recursion (Krylov) vectors and are obtained as

$$\mathbf{H}\phi_0 = \alpha_0\phi_0 + \beta_1\phi_1 \tag{5.41}$$

$$\mathbf{H}\phi_k = \beta_k\phi_{k-1} + \alpha_k\phi_k + \beta_{k+1}\phi_{k+1} \tag{5.42}$$

where ϕ_0 is the initial wave function. Thus the Hamiltonian operates on the initial wave function ϕ_0 and creates the nearest dynamical environment around ϕ_0. If the step length was zero the initial wave function would be an eigenfunction, i.e., $\beta_1 = 0$. Equations (5.41) and (5.42), together with the orthonormality conditions

$$\langle \phi_n | \phi_m \rangle = \delta_{nm} \qquad (5.43)$$

define the recursion coefficients α_k and β_k. The transformed, i.e., tridiagonal matrix representation, is obtained as

$$\tilde{\mathbf{H}} = \mathbf{T}^+\mathbf{H}\mathbf{T} = \begin{bmatrix} \alpha_0 & \beta_1 & 0 & \cdots & 0 & 0 \\ \beta_1 & \alpha_1 & \beta_2 & \cdots & 0 & 0 \\ 0 & \beta_2 & \alpha_2 & \cdots & 0 & 0 \\ \cdot & \cdot & \cdot & & \cdot & \cdot \\ \cdot & \cdot & \cdot & \cdot & & \cdot \\ \cdot & \cdot & \cdot & & \cdot & \cdot \\ 0 & 0 & 0 & \cdots & \alpha_{L-2} & \beta_{L-1} \\ 0 & 0 & 0 & \cdots & \beta_{L-1} & \alpha_{L-1} \end{bmatrix} \qquad (5.44)$$

The tridiagonal matrix is of order $L \times L$ and can easily be diagonalized. Thus the propagated wave function is obtained through

$$\phi(t + \Delta t) = \mathbf{T}\mathbf{S} \, \exp\left(-\frac{i}{\hbar} \mathbf{D}\Delta t\right) \mathbf{S}^+\mathbf{T}^+\phi(t) \qquad (5.45)$$

where $\mathbf{D}$ is a diagonal ($L \times L$) matrix containing the eigenvalues of the tridiagonal matrix; $\mathbf{T}$ is an ($N \times L$) matrix with the L recursion vectors of length N (the number of grid points); and $\mathbf{S}$ is an ($L \times L$) matrix containing the eigenvectors of the tridiagonal matrix. The number of Lanczos recursions depends upon the length of the time step and the magnitude of the Hamiltonian. But with a time step of 10^{-16} sec, 10–20 recursions are necessary for the last few recursion coefficients to be small, i.e., the recursion truncates. The rapid convergence of the recursion requires that the Hamiltonians be bounded to a region $H_{min} \leq H \leq H_{max}$ at each grid point. Thus if the potential goes to infinity in certain regions it is simply replaced by a maximum value V_{max} in these regions. The value of V_{max} depends, of course, on the actual problem under consideration and the final result should be independent of the choice.

For a given input vector, the Lanczos method will create the nearest dynamical environment around the input state. The method requires one FFT evaluation per iteration and is very stable, i.e., can be used for long time propagations.

EXERCISE

5.1. Show that the error in the split operator method is of order Δt^3. (Use the Baker-Hausdorff formula).

5.4 BOHM TRAJECTORIES

The quantum mechanical equations of motion can be formulated in a manner that, for some purposes, is more convenient than grid methods. Formally this is achieved by expressing the wave function as

$$\Psi = A(\mathbf{x}, t)\, \exp\left(\frac{i}{\hbar}\, S(\mathbf{x}, t)\right) \tag{5.46}$$

where $S(x, t)$ is the action. If we insert this in the TDSE we obtain the following set of equations

$$\frac{\partial S}{\partial t} + \frac{1}{2}\, mv^2 + V(\mathbf{x}, t) + Q(\mathbf{x}, t) = 0 \tag{5.47}$$

$$\frac{\partial P}{\partial t} + \nabla \cdot (P\mathbf{v}) = 0 \tag{5.48}$$

where $\mathbf{v} = 1/m(\nabla S)$, $P = A^2$, and

$$Q = -\frac{\hbar^2}{2m}\, \frac{\nabla^2 A}{A} \tag{5.49}$$

If $Q = 0$, Eqs. (5.47) and (5.48) are the classical equations for the action S and the probability density P. This value of Q is obtained in the classical limit. But we also see that Q has the appearance of an extra potential, the so-called quantum potential [46]. If the quantum potential is neglected ($\hbar \rightarrow 0$) the above equations reduce to those obtained classically. Thus the solution of the quantum mechanical problem is identical to the solution of the classical mechanical problem except for the addition of an extra potential—the quantum potential. Initially the trajectories are weighted with the quantum distribution function

$$P(\mathbf{x}, 0) = |\Psi(\mathbf{x}, 0)|^2 \tag{5.50}$$

The quantum ensemble of trajectories must be propagated simultaneously since they are coupled by the quantum potential. Classically, they may of course be propagated independent of each other. This similarity between classical and quantum mechanics has been used to formulate a unique definition of classical and quantum chaos in terms of a Lyapunov exponent [47]. The advantage of this approach is that the grid points (trajectories) are taken from an initially localized wave function and that only these trajectories have to be followed. In contrast to this, conventional grid methods incorporate a number of points that cover space where the wave function is not located initially.

5.5 TUNNELING OF WAVEPACKETS

Consider the one-dimensional motion of a particle with mass m on the x-axis. The Hamiltonian is given by

$$-\frac{\hbar^2}{2m}\frac{d^2}{dx^2}\psi(x) + V(x)\psi(x) = E\psi(x) \tag{5.51}$$

For some potentials this equation can be solved analytically. One example with some bearing on chemical reactivity is the Eckart potential, where (see Fig. 5.1)

$$V(x) = \frac{A}{\cosh^2(ax)} \tag{5.52}$$

The transmission factor, i.e., the probability for a particle to pass the barrier from left to right is given by (see for example ref. [48])

$$P_{\text{eck}}(k) = \frac{\cosh\left(\dfrac{2\pi k}{a}\right) - 1}{\cosh\left(\dfrac{2\pi k}{a}\right) + \cosh\left(\pi\sqrt{\dfrac{8mA}{a^2\hbar^2} - 1}\right)} \tag{5.53}$$

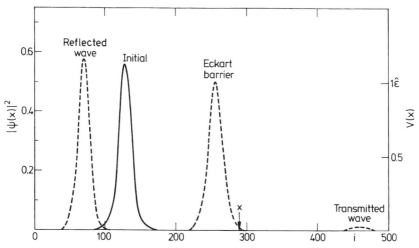

Fig. 5.1. A wavepacket is started to the left of the barrier. Part is reflected by the potential, the rest is transmitted. The energy resolution is obtained by a time-energy Fourier transform at (x).

where k is the wavenumber related to the energy by $E = \hbar^2 k^2/2m$, and where A and a are potential parameters (the barrier height and the width of the barrier). The expression was obtained solving the time-independent Schrödinger equation. A solution can also be obtained solving the time-dependent Schrödinger equation, i.e.,

$$i\hbar \, \frac{\partial \psi(t)}{\partial t} = \hat{H}\psi(t) \tag{5.54}$$

Initially the wave function is set equal to a wavepacket, i.e.,

$$\psi(x, t_0) = \left(\frac{1}{2\pi\Delta x^2} \right)^{1/4} \exp(-ik_0 x) \, \exp\left(-\frac{(x - x_0)^2}{4\Delta x^2} \right) \tag{5.55}$$

This Gaussian wavepacket is initially centered around x_0 to the left of the barrier (see Fig. 5.1). In order to avoid interference with absorbing potentials or the grid-edge, it is convenient to have a wavepacket that at first contracts and then expands. Such a wavepacket is given by

$$
\begin{aligned}
\psi(x, t_0) = &\left[2\pi\Delta x^2 \left(1 + \frac{\hbar^2 t_{\text{foc}}^2}{4m^2 \Delta x^4} \right) \right]^{-1/4} \exp(-ik_0 x) \\
&\cdot \exp\left[-\frac{(x - x_{\text{foc}} - \hbar t_{\text{foc}} k_0/m)^2}{4(\Delta x)^2 - 2i\hbar t_{\text{foc}}/m} \right]
\end{aligned} \tag{5.56}
$$

where t_{foc} is the time at which the wavepacket is focused around x_{foc} with a width Δx. The exponent may be rewritten as

$$\frac{(x - x_{\text{foc}} - \hbar k_0 t_{\text{foc}}/m)^2}{4\Delta x^2 + \hbar^2 t_{\text{foc}}^2/m^2} \left(1 + \frac{i\hbar t_{\text{foc}}}{2m\Delta x^2} \right) \tag{5.57}$$

which shows that the initial width is given by

$$\Delta x \sqrt{1 + \frac{\hbar^2 t_{\text{foc}}^2}{4m^2 \Delta x^2}} \tag{5.58}$$

It should also be noted that the momentum distribution obtained with this wavepacket is the same as for the original and is given by the Eq. (5.59) below.

If one of the time-dependent propagator methods described above is used, then the solution to the TDSE at some later time can be found. We see that the wavepacket will spread on the grid, and that a part of it be transmitted through the barrier—the remainder being reflected. In order to extract information on the transmission probability as a function of energy we first have to realize that the wavepacket (WP) corresponds not to one but actually to all energies distributed around k_0, i.e., in momentum (or k-space) we have

$$\psi_k^- = \frac{1}{\sqrt{2\pi}} \int_{-\infty}^{\infty} dx \, \exp(ikx)\psi(x, t_0)$$

$$= \left(\frac{2}{\pi}\right)^{1/4} \sqrt{\Delta x} \, \exp[i(k - k_0)x_0 - \Delta x^2(k - k_0)^2] \qquad (5.59)$$

Thus a given k-component is represented in the initial wavepacket with the weight $|\psi_k^-|^2$. In order to obtain the transmitted part of this component we can perform the space-momentum transformation at a later time t using only the transmitted part of the wavepacket, $(x > 0)$.

$$\psi_k^+ = \frac{1}{\sqrt{2\pi}} \int_{-\infty}^{\infty} dx \, \exp(-ikx)\psi(x, t) \qquad (5.60)$$

The transmission probability is then obtained as the ratio between the outgoing and the incoming fluxes

$$P_{\text{trans}}(k) = \frac{\dfrac{k_{\text{out}}}{m} |\psi_k^+|^2}{\dfrac{k_{\text{in}}}{m} |\psi_k^-|^2} \qquad (5.61)$$

For the case considered here we have $k_{\text{in}} = k_{\text{out}} = k$, but in general, e.g., for tri-atomic reactive problems, they will be different and are given by the equation

$$E = E_\alpha + \frac{\hbar^2 k_{\text{in}}^2}{2\mu_\alpha} = E_\beta + \frac{\hbar^2 k_{\text{out}}^2}{2\mu_\beta} \qquad (5.62)$$

where μ_α and μ_β are reduced masses for the fragment channels α and β, E_α and E_β, are the internal (vibrational/rotational) energies of the diatomic fragments, and E is the total energy.

The space-momentum transform requires, however, that we at time t have all the wave functions on the grid, i.e., a large grid may be required. In order to avoid reflections from the grid boundaries one can apply an imaginary absorbing potential on the last and first part of the grid. Thus in order to use the above method the wavepacket has to be represented on the nonabsorbing part of the grid. Another method that circumvents this problem is to use the time-energy rather than the space-momentum transform, i.e., we define

$$\psi_E^+ = \frac{1}{\sqrt{2\pi}} \int_{-\infty}^{\infty} dt \, \exp(iEt/\hbar)\psi(t, x^*) \tag{5.63}$$

where the "projection point" (x^*) is placed before the absorbing potential and well after the barrier. In this method we just record the wave function in each time step at the grid point x^* from, for example, $t = t_0$. The transmission factor can then be obtained as

$$P_{\text{trans}}(k) = \frac{\dfrac{\hbar^2 k_{\text{out}}^2}{m^2} \dfrac{k_{\text{out}}}{m} |\psi_E^+|^2}{\dfrac{k_{\text{in}}}{m} |\psi_k^-|^2} \tag{5.64}$$

where we have used the relation between ψ_E^+ and ψ_k^+ for a free particle

$$\psi_E^+ = \frac{m}{\hbar k} \exp(-ikx)\psi_k^+ \tag{5.65}$$

The total transmission factor can be obtained as

$$P_{\text{tot}}(k_0) = \lim_{t \text{ large}} \int_0^{x_{\max}} dx \, |\psi(x, t)|^2 \tag{5.66}$$

i.e., by "grid summation" or integration over the region to the right of the barrier (placed at $x = 0$). If part of the wave function has been absorbed by an imaginary potential placed on the grid after $x_{\max}$, we must add the time-integrated flux evaluated at $x_{\max}$

$$\frac{i\hbar}{2m} \int_0^{\infty} dt \left(\psi(x, t) \frac{\partial \psi(x, t)^*}{\partial x} - \psi(x, t)^* \frac{\partial \psi(x, t)}{\partial x} \right) \Bigg|_{x = x_{\max}} \tag{5.67}$$

to Eq. (5.66). The total transmission factor can also be obtained as

$$P_{\text{tot}}(k_0) = \int dk \, |\psi_k^-|^2 P_{\text{trans}}(k) \tag{5.68}$$

This equation can be inverted so as to obtain $P_{\text{trans}}(k)$ if the value of $P_{\text{tot}}(k_0)$ is known. This value can be obtained by grid summation, i.e., by integrating over the part of the grid to the right of the barrier (see Fig. 5.1). Hence, this method can be used when the wave functions on which the transmitted or scattered wave should be projected are either not known or complicated to determine.

Table 5.1 shows the numerical results obtained by propagating a Gaussian wavepacket through an Eckart potential. The numerical results compare well with the exact values obtained by Eq. (5.53).

For the parabolic barrier also it is possible to solve the transmission problem exactly (see for example ref. [49]). For the potential

$$V(x) = V_0 \left(1 - \frac{x^2}{x_0^2} \right) \quad \text{for } |x| < x_0$$
$$V(x) = 0 \qquad\qquad\qquad |x| > x_0$$

we have

$$P_{\text{trans}}(k) = \frac{1}{1 + \exp(2\pi(V_0 - E)/(\hbar\omega))} \tag{5.69}$$

where $\omega = \sqrt{2V_0/m}/x_0$. Table 5.2 gives the transmission probability as a function of wavenumber for a parabolic barrier with the same barrier height and half width as the Eckart potential.

TABLE 5.1. Comparison of Numerical and Exact Transmission Probabilities[a]

k (Å^{-1})	Energy kJ/mol	Numerical	Exact
15	45.374	1.301×10^{-10}	1.300×10^{-10}
16	51.626	3.009×10^{-9}	3.008×10^{-9}
17	58.280	6.953×10^{-8}	6.961×10^{-8}
18	65.339	1.611×10^{-6}	1.611×10^{-6}
19	72.800	3.727×10^{-5}	3.727×10^{-5}
20	80.665	8.617×10^{-4}	8.618×10^{-4}
21	88.933	1.957×10^{-2}	1.957×10^{-2}
22	97.604	0.3160	0.3160
23	106.68	0.9145	0.9144
24	116.16	0.9960	0.9960
P_{tot}		0.1906	0.1906

[a]Obtained using a split operator propagation method and a grid with N = 512 points. The potential parameters are A = 100 kJ/mole and a = 2.0 Å^{-1}; m = 1.0 amu, k_0 = 20 Å^{-1}, x_{min} = −20 Å, x_{max} = 20 Å and Δx = 0.2 Å. The wavepacket was propagated over 2000 times steps with Δt = 10^{-16} sec. The time-integrated flux using Eq. (5.77) is 0.1906.

TABLE 5.2. Transmission Probability Through a Parabolic Barrier[a]

k (Å^{-1})	Numerical	Exact
18	2.767×10^{-7}	2.734×10^{-7}
19	7.290×10^{-6}	7.074×10^{-6}
20	2.198×10^{-4}	2.182×10^{-4}
21	7.956×10^{-3}	7.961×10^{-3}
P_{tot}	1.087×10^{-3}	1.093×10^{-3}

[a]$V_0 = 100$ kJ/mol, $x_0 = 0.623226$ Å; m = 1.0 amu, $k_0 = 20$ Å^{-1}. The grid used in this example is the same as that given in Table 5.1

5.6 ABSORBING BOUNDARIES

In order to avoid reflection from the boundaries one can apply an imaginary potential on the grid points near x_{min} and x_{max}. The absorbing potential should be of the type

$$V_{Im}(x) = -i V_{max} f(x) \tag{5.70}$$

where $f(x)$ is a function which increases for $x \to x_{min}$ or x_{max}. The imaginary potential will kill the wave function on the grid points where it is nonzero, which is obvious from the form of the split operator propagation scheme (see Section 5.3). The function $f(x)$ should be capable of absorbing both the slow and fast energy components of the wavepacket. An a priori absorbing potential has been suggested in ref. [50]:

$$f(x) = \frac{x - x_1}{x_2 - x_1} \tag{5.71}$$

where $x_2 > x > x_1$. Then the range $(x_2 - x_1)$ where the imaginary potential is used typically account for 10% to 30% of the grid. In order to absorb slow, as well as fast, components of the wavepacket the following requirement should be fulfilled:

$$\frac{\hbar^2 k_{max}}{4m(x_2 - x_1)} \ll V_{max} \ll \frac{\hbar^2 k_{min}^3 (x_2 - x_1)}{m} \tag{5.72}$$

which leads to the following reasonable choice

$$V_{max} = \frac{\hbar^2}{2m} \sqrt{k_{max} k_{min}^3} \tag{5.73}$$

where $k_{\min} = k_0 - \Delta k$ and $k_{\max} = k_0 + \Delta k$. Here, Δk is the spread of the wavepacket in k-space

$$\Delta k = \frac{1}{2\Delta x} \tag{5.74}$$

It should also be noted that problems with imaginary absorbing potentials can occur for the time propagation with, e.g., the Lanczos scheme, due to complex eigenvalues. In such cases a simpler scheme would be to multiply the wave function with an exponential damping factor in the absorption region. Actually this is the effect that comes out of the split operator method in which the negative imaginary absorbing potential gives rise to a multiplicative factor of

$$\exp\left(-\frac{i}{\hbar} V_{\mathrm{Im}}(x)\Delta t\right) \tag{5.75}$$

Before the wavepacket is absorbed by the imaginary potential one can record the flux over a boundary (x^*) placed before the absorbing potential, i.e.,

$$
\begin{aligned}
Flux &= \frac{i\hbar}{2m}\left(\psi\,\frac{\partial\psi^*}{\partial x} - \psi^*\,\frac{\partial\psi}{\partial x}\right)\Bigg|_{x*} \\
&= \frac{\hbar}{m}\left(\mathrm{Re}\psi\,\frac{\partial\mathrm{Im}\psi}{\partial x} - \mathrm{Im}\psi\,\frac{\partial\mathrm{Re}\psi}{\partial x}\right)\Bigg|_{x*}
\end{aligned}
\tag{5.76}
$$

The time-integrated flux gives the total transmission factor (or reaction probability), i.e.,

$$P_{\mathrm{tot}}(k_0) = \int_{t_0}^{\infty} dt\,\frac{i\hbar}{2m}\left(\psi\,\frac{\partial\psi^*}{\partial x} - \psi^*\,\frac{\partial\psi}{\partial x}\right)\Bigg|_{x*} \tag{5.77}$$

As mentioned above, the insertion of an imaginary absorbing potential can be replaced by an exponential damping term in the absorption region. In this manner it is possible to avoid problems with complex eigenvalues in the time propagation scheme giving rise to exponentially growing terms. In practice the imaginary potential does not give 100% absorption, i.e., a small fraction can be reflected. Whether this is important or not depends upon the length of time the wavepacket has to be propagated, i.e., whether on the timescale of interest, one can get interference with the wave reflected from the boundaries. However, for a discussion of perfect absorbers see ref. [51].

EXERCISES

5.2. Derive Eq. (5.65) using the free particle propagator:

$$K(x,t;x_0,t_0) = \sqrt{\frac{m}{2i\pi\hbar(t-t_0)}} \exp\left[\frac{im}{2\hbar(t-t_0)}(x-x_0)^2\right] \tag{5.78}$$

and that

$$\int_0^\infty dt \, \exp(-at^2 - b/t^2) = \frac{1}{2}\sqrt{\frac{\pi}{a}} \, \exp(-2\sqrt{ab}) \tag{5.79}$$

5.3. Calculate the reaction constant for a reaction with a parabolic barrier, using

$$k(T) = \frac{1}{Q(T)h} \int_0^\infty dE \, \exp(-\beta E)P_{\text{trans}}(k) \tag{5.80}$$

5.4. Show that the wave function of Eq. (5.56), when propagated with the free particle propagator with $\Delta t = t_{\text{foc}}$, gives a Gaussian wavepacket centered around $x = x_{\text{foc}}$ with a width Δx.

6

POTENTIAL ENERGY SURFACES

Chemical reactions in the gas phase are often studied using just a single adiabatic Born-Oppenheimer potential energy surface. However nonadiabatic effects, i.e., coupling between different electronic states is an important aspect in chemistry. Thus one should be able to include such coupling whenever it is important. The wavepacket and grid methods described in the previous chapter are not restricted to a single BO surface but can readily be used for multisurface calculations. In order to incorporate nonadiabatic effects the starting point is the nonrelativistic Schrödinger equation $H\Phi = E\Phi$, where the Hamiltonian is

$$H = -\sum_a \frac{\hbar^2}{2M_a}\nabla_a^2 - \sum_j \frac{\hbar^2}{2m_e}\nabla_j^2 - \sum_a \sum_j \frac{Z_a e^2}{R_{aj}}$$

$$+ \sum_i \sum_{j<i} \frac{e^2}{R_{ij}} + \sum_a \sum_{b<a} \frac{Z_a Z_b}{R_{ab}} \qquad (6.1)$$

where a and b denote nuclei, i and j electrons. The masses of the nuclei are M_a, and that of the electrons m_e; Z_a is the charge of the nuclei; and R_{ab}, R_{aj}, and R_{ij} denote the distances between the particles. In order to facilitate the solution of the problem we invoke the so-called Born-Oppenheimer separation of the electronic and nuclear degrees of freedom. In order to obtain a solution to

$$H\Phi(\mathbf{x}, \mathbf{R}) = E\Phi(\mathbf{x}, \mathbf{R}) \qquad (6.2)$$

81

where the electronic (nuclear) coordinates are given by $\mathbf{x}(\mathbf{R})$. We utilize a Born-Oppenheimer procedure that involves five steps:

(1) Assume that the nuclei are clamped in fixed positions, therefore eliminating the nuclear kinetic energy term.
(2) Solve the electronic structure problem and obtain the electronic wave functions and corresponding energies for the given nuclear configuration.
(3) Assume that the total wave function for a given electronic state can be written as a product of an electronic and a nuclear wave function.
(4) Substitute the product wave function into Eq. (6.2).
(5) Neglect the effect of the nuclear kinetic energy operator on the electronic wave function.

6.1   THE BORN-OPPENHEIMER SEPARATION

The wave function for the electron-nuclei problem is now written as

$$\Phi(\mathbf{x}, \mathbf{R}) = \xi(\mathbf{x}; \mathbf{R})\psi(\mathbf{R}) \qquad (6.3)$$

where $\mathbf{x}$ and $\mathbf{R}$ denote the set of electron and nuclear Cartesian coordinates, respectively. We now imagine that we could solve the electronic problem first for fixed nuclei configuration. Thus the wave functions $\xi(\mathbf{x}; \mathbf{R})$ is obtained as the solution to

$$\left(-\sum_i \frac{\hbar^2}{2m_e}\nabla_i^2 + V_{eN} + V_{ee}\right)\xi(\mathbf{x}; \mathbf{R}) = E_{el}(\mathbf{R})\xi(\mathbf{x}; \mathbf{R}) \qquad (6.4)$$

where V_{eN} and V_{ee} denote the third and fourth terms in Eq. (6.1). The electronic surface $E_{el}(\mathbf{R})$ is a function of the chosen nuclear configuration. Since the equation above has several solutions corresponding to different electronic states, the surface as well as the wave function ξ should be labeled by an electronic state number, which for simplicity has been omitted here.

If we insert the product form Eq. (6.3) into the Schrödinger equation we get

$$\left(-\sum_a \frac{\hbar^2}{2M_a}\nabla_a^2 + \sum_a\sum_{b<a} \frac{Z_a Z_b}{R_{ab}} + E_{el}(\mathbf{R})\right)\psi(\mathbf{R})\xi(\mathbf{x}; \mathbf{R}) = E\psi(\mathbf{R})\xi(\mathbf{x}; \mathbf{R})$$

$$(6.5)$$

Multiplying from the left with $\xi^*(\mathbf{x}; \mathbf{R})$ and integrating over the electronic coordinates we get

$$\left(-\sum_a \frac{\hbar^2}{2M_a} \nabla_a^2 + V(\mathbf{R})\right) \psi(\mathbf{R}) = E\psi(\mathbf{R}) \tag{6.6}$$

where we have neglected the nonadiabatic coupling elements

$$-\sum_a \frac{\hbar^2}{2M_a} (2\nabla_a \psi \nabla_a \xi + \psi \nabla_a^2 \xi) \tag{6.7}$$

The magnitude of these terms is a measure of the rate of nonadiabatic transition between the different electronic levels. With this approximation the problem is reduced to the solution of the nuclear motion on a single potential energy surface

$$V(\mathbf{R}) = E_{\text{el}}(\mathbf{R}) + \sum_{a<b} \frac{Z_a Z_b}{R_{ab}} \tag{6.8}$$

We will now assume that this surface is available from electronic structure calculations or obtained by some semiempirical or empirical method.

Let us return briefly to Eq. (6.4) and rewrite it as

$$H_{\text{elec}}\xi(\mathbf{x}; \mathbf{R}) = E_{\text{el}}(\mathbf{R})\xi(\mathbf{x}; \mathbf{R}) \tag{6.9}$$

where

$$H_{\text{elec}} = -\sum_i \frac{\hbar^2}{2m_e} \nabla_i^2 + V_{eN} + V_{ee} \tag{6.10}$$

We then multiply with $\xi^*(\mathbf{x}; \mathbf{R})$ from the left and integrate over the electronic coordinates to obtain

$$E_{\text{el}}(\mathbf{R}) = \int \xi^*(\mathbf{x}; \mathbf{R}) H_{\text{elec}}\xi(\mathbf{x}; \mathbf{R}) \, d\mathbf{x} \tag{6.11}$$

We then evaluate the derivative of $E_{\text{el}}(\mathbf{R})$ with respect to a nuclear coordinate R_i

$$\frac{\partial E_{el}(\mathbf{R})}{\partial R_i} = \frac{\partial}{\partial R_i} \int \xi^*(\mathbf{x}; \mathbf{R}) H_{elec}(\mathbf{x}; \mathbf{R}) \, dx$$

$$= \int \frac{\partial \xi^*(\mathbf{x}; \mathbf{R})}{\partial R_i} H_{elec} \xi(\mathbf{x}; \mathbf{R}) \, dx + \int \xi^*(\mathbf{x}; \mathbf{R}) \frac{\partial H_{elec}}{\partial R_i} \xi(\mathbf{x}; \mathbf{R}) \, dx$$

$$+ \int \xi^*(\mathbf{x}; \mathbf{R}) H_{elec} \frac{\partial \xi(\mathbf{x}; \mathbf{R})}{\partial R_i} \, dx \tag{6.12}$$

Since the sum of the first and third integrals is zero we find that

$$\frac{\partial E_{el}(\mathbf{R})}{\partial R_i} = \int \xi^*(\mathbf{x}; \mathbf{R}) \frac{\partial H_{elec}}{\partial R_i} \xi(\mathbf{x}; \mathbf{R}) \, dx \tag{6.13}$$

which is known as the Hellmann-Feynman theorem. It shows that within the Born-Oppenheimer approximation the forces on the nuclei are given with respect to the derivative of H_{elec} averaged over the electronic wave function.

6.2   SURFACE CHARACTERISTICS

The potential energy surface $V(\mathbf{R})$ for a molecule composed of N atoms is a function of $3N - 6$ internuclear internal coordinates, e.g., three variables for a triatomic, six for a tetra-atomic system, etc. The calculation and characterization of this multidimensional function presents a major challenge in theoretical chemistry. Although the potential is a global function of the coordinates, some features and regions are more important than others. Regions of interest for spectroscopy and for the study of chemical reactions are, for example, minimia and saddle points of first order. Minima correspond to stable forms of the molecule: the global minimum to the most stable form, and local minima to stable forms of higher energy. A saddle point of first order is the point of highest energy along the minimum energy path between two minima, a mountain path connecting two valleys—the reactant and the product valleys. At the saddle point and the minima, the gradient vanishes, i.e.,

$$g_k = \frac{\partial V}{\partial R_k} = 0 \tag{6.14}$$

where $k = 1, 2, \ldots, 3N$, and R_k denotes the cartesian nuclear coordinates.

At the critical point, there is no force acting on the atoms in the molecule. For further characterization we also introduce the so-called Hessian matrix, defined by the second derivative or force constant matrix

$$H_{kl} = \frac{\partial^2 V}{\partial R_k\, \partial R_l} \tag{6.15}$$

At a minimum, the eigenvalues of this Hessian matrix are all positive (the matrix is positive definite). At a saddle point of first order, one of the eigenvalues is negative, i.e., there is a maximum in one dimension and a minimum in $3N - 7$ dimensions. Saddle points of higher order will then have more negative eigenvalues.

In order to find the critical points one can use the so-called Newton-Raphson method, in which the potential is expanded at a point close to the critical point R_c,

$$V(\mathbf{R}) = V(\mathbf{R}_0) + \sum_{k=1}^{3N-6} g_k(\mathbf{R}_0)\,\Delta R_k + \frac{1}{2} \sum_k \sum_l H_{kl}(\mathbf{R}_0)\,\Delta R_k \Delta R_l \tag{6.16}$$

where

$$\Delta R_k = R_k - R_{0k} \tag{6.17}$$

In matrix vector notation we write

$$V(\mathbf{R}) = V(\mathbf{R}_0) + \mathbf{g}^T \Delta \mathbf{R} + \tfrac{1}{2}\Delta \mathbf{R}^T \mathbf{H} \Delta \mathbf{R} \tag{6.18}$$

Or for the derivative near the critical point

$$\mathbf{g}(\mathbf{R}) = \mathbf{g}(\mathbf{R}_0) + \mathbf{H}(\mathbf{R}_0)\,\Delta \mathbf{R} \tag{6.19}$$

and hence by the definition of a stationary point we get

$$\Delta \mathbf{R} = -\mathbf{H}(\mathbf{R}_0)^{-1}\mathbf{g}(\mathbf{R}_0) \tag{6.20}$$

This equation defines the Newton-Raphson step in the direction of the minimum. Only if the potential was an exact quadratic function over the entire region $\mathbf{R}_0$ to $\mathbf{R}_c$ would it be possible to reach the critical point in one step. This is usually not the case, hence the derivative and the Hessian have to be calculated at the new point and the procedure has to be repeated until convergence.

As mentioned above the potential energy surface (PES) is a function of $3N - 6$ internal coordinates describing the relative positions of the nuclei. Thus, at present, the complete surface is known to "chemical accuracy" only for very few triatomic systems. By chemical accuracy we mean the accuracy needed to predict the rate constant at room temperature with an accuracy better than 30% to 50%.

This requires about 1 kJ/mol (0.2 kcal/mol) estimates of the barrier height. For the prediction of differential cross sections and scattering resonances in agreement with the best experimental resolution, even higher accuracy is needed, and in addition not just around the saddle point. Although significant progress has been made in the "exact," i.e., quantum mechanical treatment of reactive scattering processes for tri- and tetra-atomic reactions [53] the requirements mentioned above give unfavorable odds when pursuing the *ab initio* quantum chemical plus dynamical treatment of other than a few benchmark systems. We shall therefore concentrate on methods that in various ways, reduce the requirements for complete information on the PES or a complete dynamical description. These methods, which we think will be useful for treating a more general class of problems, are the reaction path, the variational transition state, the flux–flux correlation method, and various statistical or phase space theories. These methods will be useful for treating larger systems, reactions in liquids on solids, etc. For example, the reaction path method requires only the gradient and the Hessian matrix along the reaction path, while the flux–flux and the transition state methods require information only around the saddle point.

7

THE REACTION PATH METHOD

In order to determine the reaction path one starts at the saddle point, determined by the Newton-Raphson technique described in the previous chapter, and follows the path of steepest descent to the reactant and product valleys. The steepest descent path takes steps in the direction that leads to a maximal decrease in potential energy. Thus we wish to maximize

$$\Delta V(\mathbf{R}) = \sum g_k \Delta R_k \tag{7.1}$$

subject to the constraint that $\Delta \mathbf{R}^2$ is fixed. Let us introduce the functional

$$I(\mathbf{R}) = \sum_k g_k \Delta R_k + \lambda \sum_k \Delta R_k^2 \tag{7.2}$$

where λ is a Lagrange multiplier. The extremum condition for ΔR_k then becomes

$$\Delta R_k = \frac{-1}{2\lambda} g_k \tag{7.3}$$

i.e., the steepest descent path is proportional to the negative gradient of the energy in that direction.

The intrinsic reaction path (IRP) is defined as the steepest descent path in mass-weighted coordinates. We now introduce the mass-weighted Cartesian position coordinates with notation, such that

$$x_i = \sqrt{m_j}\, X_{j,\alpha} \tag{7.4}$$

where $i = 1, 2, 3$ for $j = 1$, and $\alpha = $ x, y, z, etc. The symbol $X_{j,\alpha}$ is the position coordinates of atom j in direction α. Along the IRP we have

$$\frac{d\mathbf{x}(s)}{ds} = -\frac{\nabla V}{|\nabla V|} \tag{7.5}$$

where s is the IRP arc length along the path, i.e., it is an integral over displacements ds defined as

$$ds = \pm \sqrt{\sum_{i=1}^{3N} (dx_i)^2} \tag{7.6}$$

where, by convention, the $(-)$ sign refers to the path from the saddle point to reactants, and the $(+)$ sign for the path to products. Note that the unit of s is $\sqrt{amu}$Å.

We notice that the IRP can be defined by the equation

$$\frac{d\mathbf{x}(t)}{dt} = -\nabla V(\mathbf{x}) \tag{7.7}$$

where t is a parameter. We now introduce a local quadratic expansion around the point $\mathbf{x}_0$, i.e.,

$$V(\mathbf{x}) = V(\mathbf{x}_0) + \nabla V(\mathbf{x}_0) \cdot (\mathbf{x} - \mathbf{x}_0) + \tfrac{1}{2}(\mathbf{x} - \mathbf{x}_0)^T \cdot \mathbf{H}(\mathbf{x}_0) \cdot (\mathbf{x} - \mathbf{x}_0) \tag{7.8}$$

From the two above equations we get

$$\frac{d}{dt}(\mathbf{x} - \mathbf{x}_0) = -\nabla V(\mathbf{x}_0) - \mathbf{H}(\mathbf{x}_0)(\mathbf{x} - \mathbf{x}_0) \tag{7.9}$$

This equation has the solution

$$\mathbf{x}(t) = \mathbf{x}_0 - \sum_{i=1}^{3N} \mathbf{T}_i^T \cdot \nabla V(\mathbf{x}_0)[(\exp(\lambda_i(t - t_0)) - 1)/\lambda_i]\mathbf{T}_i \tag{7.10}$$

where

$$-\mathbf{HT}_i = \lambda_i \mathbf{T}_i \qquad (7.11)$$

i.e., $\mathbf{T}_i$ are eigenvectors and λ_i the corresponding eigenvalues to the Hessian matrix evaluated in mass-weighted coordinates. Thus the IRP is found by starting at the saddle point and using the above scheme in the direction of products and reactants. At each point we need to calculate the gradient and the Hessian. The value of s is updated along the path using Eq. (7.6). Note that the above procedure fails at the saddle point itself (the gradient being zero). Thus the first small step is taken in the positive or negative direction of the vector corresponding to the imaginary frequency, which is found by diagonalizing the Hessian at the saddle point.

7.1 REACTION PATH CONSTRAINTS

Consider a rigid body rotating about a fixed axis with a constant angular velocity, ω, in radians per second. This rotation can be described by a vector $\boldsymbol{\omega}$ with length ω and a direction parallel to the axis of rotation. A point P not on the rotation axis will then have a linear velocity given by

$$\mathbf{v} = \omega \times \mathbf{r} \qquad (7.12)$$

where $\mathbf{r}$ is the radius vector from the point P to a fixed point O on the axis of rotation. If the point has a mass m the momentum is

$$m\mathbf{v} = m(\omega \times \mathbf{r}) \qquad (7.13)$$

and the angular momentum about the point O is (see Fig. 7.1)

$$\mathbf{M} = \mathbf{r} \times m\mathbf{v} = m[\mathbf{r} \times (\omega \times \mathbf{r})] \qquad (7.14)$$

Considering now the motion of a molecule consisting of N atoms. We introduce the position vector of atom i in a space-fixed coordinate system

$$\mathbf{r}_{0i} = \mathbf{r}' + \mathbf{r}_i \qquad (7.15)$$

where the vector $\mathbf{r}'$ denotes the origin of the coordinate system $O'_{x'y'z'}$ located at the center of mass of the molecule, and $\mathbf{r}_i$ is the position vector of atom i in the coordinate system $(x'y'z')$. The motion of an atom i relative to the center of mass can be thought of as consisting of two contributions: a rotational motion around the center of mass connected to the rotation of a rigid framework where the atoms have their equilibrium positions, and an oscillatory or displacement

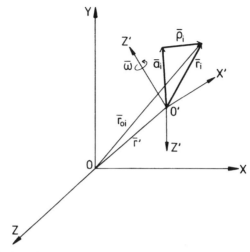

Fig. 7.1. The velocity of a particle (i) in a space-fixed coordinate system is divided in center of mass, rotational motion, and vibrational displacement contributions.

motion from this position. Thus the position vector $\mathbf{r}_i$ can be written as

$$\mathbf{r}_i = \mathbf{a}_i + \rho_i \tag{7.16}$$

where $\mathbf{a}_i$ denotes the equilibrium position vector, and ρ_i the displacement vector. Considering now the velocity of an atom i we have, in the space-fixed coordinate system (xyz) (see Figs. 7.1 and 7.2):

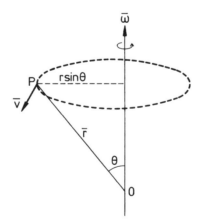

Fig. 7.2. Angular velocity ω of a rigid body gives rise to a linear velocity $\mathbf{v}$ of a point P.

$$\mathbf{v}_{0i} = \mathbf{v}' + (\omega \times \mathbf{r}_i) + \mathbf{v}_i \tag{7.17}$$

where the first term is the velocity of the center of mass, the second the linear velocity of the atom i due to the rotation of the molecular around the origin $O'_{x'y'z'}$, and the last term is the translational velocity of the atom relative to the $O'_{x'y'z'}$, i.e., $\mathbf{v}_i = d\mathbf{r}_i/dt$. The motion is subject to six constraints, namely that

$$\sum_i m_i \mathbf{v}_i = \mathbf{0} \tag{7.18}$$

These three equations fix the center of mass at the origin $O'_{x'y'z'}$. The other three constraints are due to Eckart and are therefore called the Eckart condition [55]. The Eckart condition states that if atom i is in its equilibrium position, where $\mathbf{r}_i = \mathbf{a}_i$, then there should be no additional angular momentum relative to $O'_{x'y'z'}$.

Considering the angular momentum we have

$$\mathbf{M} = \sum_i m_i [\mathbf{r}_i \times (\omega \times \mathbf{r}_i)] + \sum_i m_i \mathbf{r}_i \times \mathbf{v}_i \tag{7.19}$$

and setting $\mathbf{r}_i = \mathbf{a}_i$ we must (due to this constraint) require that

$$\sum_i m_i \mathbf{a}_i \times \mathbf{v}_i = \mathbf{0} \tag{7.20}$$

since the first term is then the angular momentum of a rigid body rotating around an axis through the orgin $O'_{x'y'z'}$ with an angular velocity ω. These six constraints leave us with $3N - 6$ degrees of freedom for the displacement from the reference frame defined by the vectors $\mathbf{a}_i$. The complete motion of the $3N$ atoms has then been divided into three parts: translational motion of the center of mass, rotational motion of a rigid frame, and displacement (vibrational) motion relative to the rotating frame.

This analysis is relevant whenever we are at a stationary point corresponding to a minimum. However, at a saddle point and along the IRP the motion (in s) along the reaction path is not of small amplitude and hence must be treated differently. Thus we are left with not $3N - 6$ but only $3N - 7$ vibrational-like motions, plus one translational motion along the reaction path. Formally this is done by introducing yet another constraint, namely that the vibrational displacement ρ_i is perpendicular to a vector along the reaction path (see Fig. 7.3). The reference vectors $\mathbf{a}_i$ become dependent on the reaction path parameter s. A vector along the path is $d\mathbf{a}_i/ds$ and the reaction path constraint is

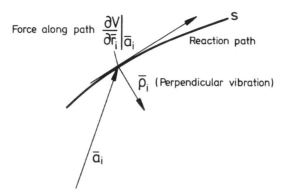

Fig. 7.3. In the reaction path method the path is defined through the constraint that a vector along the path should be perpendicular to the displacement vector ρ.

$$\sum_i m_i \frac{d\mathbf{a}_i}{ds} \cdot \rho_i = 0 \qquad (7.21)$$

where $d\mathbf{a}_i/ds$ is a vector along the reaction path for atom i, and, as we have seen above,

$$\frac{d\mathbf{a}_i}{ds} = -\text{const} \left. \frac{\partial V}{\partial \mathbf{r}_i} \right|_{\mathbf{r}_i = \mathbf{a}_i} \qquad (7.22)$$

By introducing this seventh constraint it is possible to derive a reaction path Hamiltonian [57, 58]. The result is

$$H = \frac{p_s^2}{2\tilde{N}} + H_{\text{rot}} + H_{\text{vib}} + V_0(s) - \frac{p_s}{\tilde{N}} \sum_{kk'} P_k Q_{k'} B_{kk'} \qquad (7.23)$$

where

$$\tilde{N} = 1 + 2 \sum_k B_{kF}(s) Q_k + \sum_{kk'} \tilde{C}_{kk'}(s) Q_k Q_{k'} \qquad (7.24)$$

where Q_k are normal mode coordinates for the perpendicular vibration $k = 1$, $2, \ldots, 3N - 7$, and P_k the corresponding momenta. The symbol F denotes the reaction path degree of freedom. The terms B_{kF} and $\tilde{C}_{kk'}$ represent Coriolis-like coupling between the reaction path motion F, the vibrational mode k, and two vibrational modes k and k', respectively. In addition, p_s is the momentum for the reaction path motion, H_{rot} the Hamiltonian for the overall rotational motion of the molecule, $V_0(s)$ the potential along the steepest descent path, and

$$H_{\text{vib}} = \frac{1}{2} \sum_{k=1}^{3N-7} (P_k^2 + \omega_k(s)^2 Q_k^2) \qquad (7.25)$$

By expanding the factor $\tilde{N}$ to second order in the vibrational coordinates Q_k, it is possible to bring the Hamiltonian into a form for which a quantum mechanical treatment within the second quantization method can be performed. Although the reaction path methodology is of course not restricted to a second order truncation in the normal mode coordinates, this approximation makes the formulation of the method especially compact.

In order to use the reaction path Hamiltonian theory in calculations of absolute cross sections, it is necessary to relate the original phase space of the reactants to that available for the system when governed by reaction path dynamics. The reaction path Hamiltonian is the Hamiltonian for a "supermolecule" consisting of the two reactant molecules forming a loose complex, the geometry of which is defined by following the reaction path from the transition state to the reactant channel. Since this geometry represents only one particular collision out of many possible encounters, it is necessary to introduce a phase space conversion factor. It has been suggested [56] that the conversion or kinematic factor, which should be introduced, is one which ensures that the reaction path approach, when the transition state approximation (no recrossing of a dividing surface) is introduced in the reaction path method, gives the same result for the reaction rate as that obtained from transition state theory. The transition state limit is thereby contained in the reaction path approach. However, the difference is that reaction path theory defines a Hamiltonian that can be used for a dynamical evaluation of the reaction rate constant, so that effects such as recrossing and tunneling can be incorporated. Hence the reaction path method retains the collisional or "dynamical point of view," whereas transition state theory adopts equilibrium statistical concepts.

Considering, for example, the collision of two diatomic molecules with initial vibrational quantum numbers v_1 and v_2, we obtain for the rate constant [56]

$$k_{v_1 v_2 \rightarrow v_1' v_2'}(T) = \sqrt{\frac{8kT}{\pi\mu}} \; \frac{\pi\hbar^2}{2\mu kT} \; \frac{\sigma}{Q_{\text{rot}}} \sum_J (2J+1) \sum_{K=-J}^{J} \int_0^{\infty} d(\beta E)$$

$$\cdot \exp(-\beta E) P_{\text{react}}(E, J, K) \qquad (7.26)$$

where we have assumed that the reaction path complex is a symmetric top with principal moments of inertia $I_{xx} \sim I_{yy}$. Consequently we have introduced the total angular momentum J and its projection on a body-fixed axis K. The term Q_{rot} is the rotational partition function of the reactants, J the total angular momentum, E the energy available to the complex, σ a symmetry factor (see Chapter 8), and μ the reduced mass of the diatomic system. The reaction probability P_{react} is in the reaction path formulation found by integrating the equations of motion numerically. In transition state theory, the flux over a dividing surface is computed by phase space or statistical means. A particularly compact formalism can be obtained by treating the perpendicular vibrational motions quantally within the second quantization approach, retaining a classical description of the rotational motion combined with either a classical (high energies) treatment of the motion along the reaction path or, for energies below the barrier, a quantum treatment of this motion using the wavepacket propagation technique (see [54]).

8

VARIATIONAL TRANSITION STATE THEORY

Variational transition state theory (VTST) can be obtained from reaction path theory if one introduces the transition state assumption, i.e., that of no recrossing over a dividing line, though the theory allows for a variation of the transition state dividing line. In order to obtain an expression for the rate constant we calculate the microcanonical reaction rate $k(E)$ and use that the bulk rate constant is given by

$$k(T) = \frac{\int dE \; k(E)\rho(E)\exp(-E/kT)}{Q(T)} \tag{8.1}$$

where $Q(T) = \int dE \; \rho_{reactant} \exp(-E/kT)$ is the partition function for the reactants and $\rho(E)$ the density of states at the energy E.

In reaction path theory the reactant channel is defined by a point along the reaction path, where the complex, consisting of N_a atoms, is separated in non-interacting fragments. Thus if we stay within the reaction path picture from reactants to products the formulation resembles that for unimolecular reactions. However, the goal of the theory is to predict bimolecular reaction rates, and hence the reaction partition functions should include the rotational, vibrational, and translational degrees of freedom of the separated fragments [see Eq. (8.15) below].

The microcanonical rate constant $k(E)$ can be obtained by calculating the number of states passing a dividing line (see Fig. 8.1) along the reaction path at $s = s_0$ per unit time. Thus

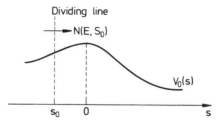

Fig. 8.1. Number of states passing the dividing line s_0 per unit time.

$$k(E) = \frac{\int ds \int dp_s \int d(\{Q_k\}) \int d(\{p_k\})\delta(H - E)\delta(s - s_0)\dot{s}\theta(\dot{s})}{\int ds \int dp_s \int d(\{Q_s\}) \int d(\{p_k\})\delta(H - E)} \tag{8.2}$$

Note that we have assumed that all phase space points have the same weight. For a system undergoing a chemical reaction, the uniform distribution of phase space points is justified by a strong coupling assumption, an ergodic hypothesis, i.e., it is assumed that the system has had enough time to distribute itself in phase space so that all points are equally probable. This assumption is of a statistical nature and underlines microscopic theories such as unimolecular reactions and transition state theory. However, in order to obtain the transition state result we must make yet another approximation, namely that there is no recrossing of the transition state dividing line. A correction to take care of this latter point deals with unique ways of defining the dividing line/surface or with obtaining the transmission factor correction, which will be discussed later (Section 8.2).

With the above assumption we now wish to find the number of states passing a dividing line along the reaction path where $(s = s_0)$ (see Fig. 8.1) subject to the constraint that the energy is equal to E. The reaction path Hamiltonian can be written (neglecting the rotational motion) as

$$H(s, p_s, \{Q_k\}, \{p_k\}) = \frac{p_s^2}{2\tilde{N}} + V_0(s) + \frac{1}{2}\sum_k (p_k^2 + \omega_k^2 Q_k^2)$$
$$- \frac{p_s}{\tilde{N}}\sum_{kk'} p_k Q_{k'} B_{kk'} \tag{8.3}$$

where $\tilde{N}$ is as given in Eq. (7.24).

In order to find the one-way flux we have introduced the function

$$\theta(\dot{s}) = \begin{cases} 1 & \text{if } \dot{s} > 0 \\ 0 & \text{if } \dot{s} < 0 \end{cases} \tag{8.4}$$

where $\dot{s}$ is the velocity with which these states pass s_0. Introducing that the density of states $\rho(E)$ is given by

$$\rho(E) = \frac{1}{h^n} \int d(\{p_q\}) \int d(\{q\}) \, \delta(H - E) \tag{8.5}$$

where $d(\{q\}) = dq_1, \ldots, dq_n$, n being the dimension of the system, and h Planck's constant, we notice that the expression for $k(E)$ can be written as

$$k(E) = \frac{\int ds \int d\tilde{p}_s \, \delta(s - s_0) \tilde{p}_s \theta(\dot{s}) \rho^{\#}(E - V_0(s) - \frac{1}{2} \tilde{p}_s^2)}{h\rho(E)} \tag{8.6}$$

where we have introduced

$$\tilde{p}_s = \frac{1}{\sqrt{\tilde{N}}} \left(p_s - \sum_{kk'} p_k Q_{k'} B_{kk'} \right) \tag{8.7}$$

and used that $\tilde{p}_s \, d\tilde{p}_s = \dot{s} \, dp_s$.

We can evaluate $\rho^{\#}$ (the density of states at a point along the reaction path) by using that

$$\rho^{\#} = \frac{dN^{\#}}{dE} \tag{8.8}$$

where $N^{\#}$ is the number of states with energy less than or equal to $E - V_0(s) - \frac{1}{2} \tilde{p}_s^2$, i.e.,

$$N^{\#}\left(E - V_0(s) - \frac{1}{2} \, \tilde{p}_s^2 \right)$$

$$= \frac{1}{h^M} \int d(\{p_k\}) \int d(\{Q_k\})\big|_{[(1/2) \sum (p_k^2 + \omega_k^2 Q_k^2) \leq E - V_0(s) - (1/2)\tilde{p}_s^2]}$$

$$= \frac{1}{h^M M!} \left(E - V_0(s) - \frac{1}{2} \, \tilde{p}_s^2 \right)^M \prod_{k=1}^{M} \frac{2\pi}{\omega_k} \tag{8.9}$$

where M is the number of vibrational degrees of freedom $3N_a - 7$. From the above expressions we obtain

$$\rho^{\#} = \frac{1}{(M-1)!} \frac{(E - V_0(s) - \frac{1}{2} \tilde{p}_s^2)^{M-1}}{\Pi_{k=1}^{M} \hbar \omega_k} \tag{8.10}$$

In order to perform the integration over $\tilde{p}_s$ we have to assume $\theta(p_s) = \theta(\tilde{p}_s)$. But in practice this would correspond to moving the dividing surface to a point along the reaction path, where $\tilde{p}_s = 0$. Thus we have

$$\frac{1}{(M-1)!} \int_0^{\tilde{p}_s^{\max}} d\tilde{p}_s \, \tilde{p}_s \, \frac{(E - V_0(s_0) - \frac{1}{2}\tilde{p}_s^2)}{\Pi \hbar \omega_k(s_0)} = \frac{(E - V_0(s_0))^M}{M! \Pi_{k=1}^M \hbar \omega_k(s_0)} \quad (8.11)$$

where $\tilde{p}_s^{\max}$ is defined by $E - V_0(s_0) - \frac{1}{2}(\tilde{p}_s^{\max})^2 = 0$. Thus we obtain finally

$$k(E, s_0) = \frac{1}{h\rho(E)} \frac{(E - V_0(s_0))^M}{M! \Pi_{k=1}^M \hbar \omega_k(s_0)} \quad (8.12)$$

which, when inserted in the expression for the rate constant, gives

$$k(T, s) = \frac{1}{hQ(T)} \frac{1}{M! \Pi \hbar \omega_k(s)} \int dE(E - V_0(s))^M \exp(-E/kT)$$

$$= \frac{kT}{h} \frac{1}{Q(T)} \exp(-V_0(s)/kT) Q_{\text{vib}}(s, T) \quad (8.13)$$

where we have replaced s_0 by s and introduced the vibrational partition function

$$Q_{\text{vib}} \sim \prod_{k=1}^M \frac{kT}{\hbar \omega_k} \quad (8.14)$$

i.e., we have obtained the transition state result. Thus we see that the total reaction rate constant is independent of the Coriolis coupling terms $B_{kk'}$. These terms play a role only if state-resolved rate constants are wanted, i.e., they determine the final state distribution.

If the rotational motion of the reaction path "complex" is included, we obtain an expression similar to Eq. (8.13), though with an additional partition function $Q_{\text{rot}}(s)$. By doing this we obtain the usual transition state expression just evaluated, at a point s along the reaction path. The rate constant then becomes

$$k(T, s) = \frac{kT}{h} \sigma \frac{Q_{\text{rot}}(s) Q_{\text{vib}}(s)}{Q_{\text{rot}}^R Q_{\text{vib}}^R Q_t^{\text{rel}}} \exp(-\beta V_0(s)) \quad (8.15)$$

where σ is a symmetry factor counting the number of equivalent reaction paths, and Q_{rot}^R and Q_{vib}^R are the partition functions for the rotational and vibrational degrees of freedom of the reactants. The term Q_t^{rel} is the translational partition function per unit volume for the relative translational motion. We note, that the rate constant will depend on where along the reaction path it is evaluated. The reason for this is of course that frequencies, moments of inertia, and the potential $V_0(s)$ depend on s. The variational transition state rate constant is now obtained using the value of $k(T,s)$ evaluated where the flux from reactants to products is at a minimum, i.e.,

$$k(T)^{VTST} = \min_s \ k(T,s) \tag{8.16}$$

In the case of several minima separated by maxima, we introduce the following definition of an optimized transition state rate

$$k_{otst}(T) = \left[\sum_{p=1,3\ldots} k_{\min}(T,s_p)^{-1} - \sum_{p=2,4\ldots} k_{\max}(T,s_p)^{-1} \right]^{-1} \tag{8.17}$$

where p odd denotes the local minima, and p even the maxima along the reaction path.

Usually the transition state is taken to be localized at the top of the barrier along the reaction path, but in principle the dividing surface is arbitrary, i.e., the position can be varied such that the equilibrium flux through the surface equals the reaction rate constant. If such a surface could be defined, the transition state theory would be exact. However, for practical purposes the transition state rate constant will always be an upper bound to the rate constant. Variational transition state theory therefore defines the dividing surface as the surface where the flux minimizes the rate constant [59].

EXERCISES

8.1. Find the number of states associated with a harmonic oscillator having an energy less than or equal to E. Use the relation $H(x,p_x) = p_x^2/2m + \frac{1}{2}kx^2$ and that

$$\iint_{x_1^2 + \ldots x_n^2 \leq R^2} dx_1, \ldots, dx_n = \frac{\sqrt{\pi^n}}{\Gamma\left(\dfrac{n}{2} + 1\right)} R^n \tag{8.18}$$

8.2. Consider a prototype S_N2 reaction [see Eq. (8.19)] at $T = 300$ K.

a. Calculate the transition state rate constant for the S_N2 reaction

$$Cl^- + CH_3Cl \rightarrow ClCH_3 + Cl^- \tag{8.19}$$

as a function of the parameter s, the distance along the reaction path. Note that $s = 0$ corresponds to the saddle point.

b. Find the variational transition state rate constant for the reaction.

c. Estimate from the second derivative d^2V/ds^2 the Wigner tunneling correction factor

$$1 + \frac{1}{24}\left(\frac{\hbar|\omega^*|}{kT}\right)^2 \tag{8.20}$$

d. Consider the complex

$$Cl^- \cdots CH_3 - Cl \tag{8.21}$$

at $s = -4.5\sqrt{amu}Å$. Collisions with a structureless molecule with mass 18 amu excite the complex, and the following unimolecular reaction occurs:

$$M + Cl^- \cdots CH_3Cl\ (s = -4.5) \rightarrow M + Cl \cdots CH_3 \cdots Cl(s = 0)$$

$$\rightarrow ClCH_3 + Cl^- + M \tag{8.22}$$

Calculate the high pressure unimolecular reaction rate k^∞_{unit}. Use Tables 8.1 and 8.2.

TABLE 8.1. Molecular Parameters as a Function of the Distance s Along the Reaction Path for the Reaction $Cl^- + CH_3Cl \rightarrow ClCH_3 + Cl^-$

s $\sqrt{amu}$ Å	$V_0(s)$ (kJ/mol)	$\frac{d^2V}{ds^2}$ (kJ/[amu mol Å²])	I^e_{xx}	I^e_{yy} (amu Å²)	I^e_{zz}
−4.50	−46.5	6.56	3.09	503.0	503.0
−0.30	12.3	−26.0	3.44	402.0	402.0
−0.20	13.7	−39.0	3.45	404.0	404.0
−0.10	14.6	−62.4	3.46	405.0	405.0
0.00	14.9	−66.6	3.46	405.0	405.0
0.10	14.6	−62.4	3.46	405.0	405.0
0.20	13.7	−39.0	3.45	404.0	404.0
0.30	12.3	−26.0	3.44	402.0	402.0

$I^e_{\alpha\alpha}(\alpha = x, y, z)$ are the principal moments of inertia.

TABLE 8.2. Vibrational Frequencies^a

s	E	A	E	A	E	E	A
−4.50	0.216	1.06	1.47	2.45	2.61	5.78	5.70
−0.30	0.387	0.638	2.00	2.00	2.61	6.17	6.47
−0.20	0.370	0.518	1.96	1.95	2.60	6.17	6.47
−0.10	0.357	0.409	1.92	1.90	2.60	6.17	6.47
0.00	0.355	0.377	1.91	1.89	2.60	6.17	6.46
0.10	0.357	0.409	1.92	1.90	2.60	6.17	6.47
0.20	0.371	0.518	1.96	1.95	2.60	6.17	6.47
0.30	0.387	0.638	2.00	2.00	2.61	6.17	6.47

$^a\omega_k$ in units of 10^{14} sec^{-1}; E denotes doubly and A singly degenerate frequencies; s is the distance along the reaction path.

8.1 TUNNELING

If the dynamics defined by the reaction path Hamiltonian are solved using classical mechanics it is necessary to introduce tunneling correction factors for energies below the barrier. In transition state theory both below and above barrier reflection or tunneling corrections are, in principle, needed. The first correction should allow for the fact that the TST gives an upper bound to the rate constant at energies above the barrier and the second for neglecting quantum tunneling through the barrier. The simplest tunneling factor often used to correct the TST value for the rate constant is that due to Wigner [60], i.e.,

$$\kappa \sim 1 + \frac{1}{24} \frac{(\hbar \omega^{\#})^2}{(kT)^2} \tag{8.23}$$

where $\omega^{\#}$ is the tunneling frequency obtained from the second derivative of the potential at the saddle point. In semiclassical theory, the tunneling factor for one-dimensional tunneling through a barrier can be evaluated as [61]

$$P_{\text{tunn}} = \frac{1}{1 + \exp\left(\dfrac{2}{\hbar} I(s_1, s_2)\right)} \tag{8.24}$$

i.e., by evaluating the integral

$$I(s_1, s_2) = \text{Im} \int_{s_1}^{s_2} ds\, p_s \tag{8.25}$$

along the reaction coordinate between the "turning" points s_1 and s_2, where

$p_s(s_i) = 0$. In the simplest case the momentum along the reaction path is obtained from

$$\frac{p_s^2}{2m} = E - V_0(s) \tag{8.26}$$

However, this formula does not include the effect of curvature on the reaction path. The reason for including the curvature is to try to correct for the contribution from the "corner cutting" trajectories, i.e., that the tunneling does not necessarily follow the minimum energy path [62]. The tunneling probability along the shortest path may exceed the one obtained along the minimum energy path although the potential (the barrier) is higher. Marcus and Coltrin [62] introduced a vibrational adiabatic approximation and tunneling along the path with maximum vibrational amplitude in the motion perpendicular to the reaction path; $Q_k(\text{max})$ is defined by

$$\tfrac{1}{2}\omega_k^2(s)Q_k^2(\text{max}) = \hbar\omega_k(n_k + \tfrac{1}{2}) \tag{8.27}$$

Thus the shorter path starts at the outer turning point for the vibrational motion and cuts the corner (see Fig. 8.2).

In the reaction path formulation it is possible to include the effect of curvature using Eq. (7.23) to define p_s [56, 64]. In the simplest application of the reaction path Hamiltonian, the vibrationally adiabatic approximation is made, i.e., no coupling between the various vibrational modes is allowed during the tunneling process. Neglecting the quadratic terms and the rotational part of the Hamiltonian we obtain from Eq. (7.23)

$$p_s = \sqrt{2\tilde{N}(s)\left(E - V_0(s) - \sum_k \hbar\omega_k(s)\left(n_k + \frac{1}{2}\right)\right)} \tag{8.28}$$

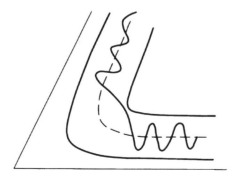

Fig. 8.2. Minimum energy path (dashed line) and a corner-cutting path (full line).

where $\tilde{N}(s) \sim 1 + 2 \sum_k B_{kF}Q_k$ acts as an effective mass parameter. We now introduce $Q_k = \sin(\phi_k)\sqrt{\hbar(2n_k + 1)/\omega_k(s)}$, where ϕ_k is a vibrational phase factor. Evaluation of the correction, from the vibrational degrees of freedom for $s = 0$ only, then gives

$$I(s_1, s_2) = \left\{ \pi \left[V_0(0) - E_{\text{kin}} + \sum_k \hbar(\omega_k(0) - \omega_k(-\infty)) \left(n_k + \frac{1}{2} \right) \right] \right\} \bigg/ \omega^{\#}$$

(8.29)

where

$$\omega^{\#} = \sqrt{ -\frac{d^2 V_0(s)}{ds^2} \bigg|_{s=0} \bigg/ \tilde{N}(0) }$$

(8.30)

We note that the tunneling will increase if $\omega_k(0) < \omega_k(-\infty)$. The reaction path approach usually includes just the terms up to second order in the modes perpendicular to the reaction path, although a more realistic evaluation of the tunneling correction factor also includes the anharmonic corrections. Comparison between exact calculations and VTST results including tunneling correction factors show [65] that such corrections can be substantial.

EXERCISES

8.3. In order to obtain an expression for the transmission factor, use the semi-classical connection formula [36]

$$\begin{pmatrix} R_+ \\ R_- \end{pmatrix} = \begin{pmatrix} \sqrt{2 + \kappa^2}\,\exp(-i\phi) & -i\kappa \\ i\kappa & \sqrt{1 + \kappa^2}\,\exp(i\phi) \end{pmatrix} \begin{pmatrix} L_+ \\ L_- \end{pmatrix}$$

(8.31)

where

$$\kappa = \exp(-\pi\epsilon)$$

(8.32)

$$\phi = \arg \Gamma\left(\frac{1}{2} + i\epsilon \right) - \epsilon \ln|\epsilon| + \epsilon$$

(8.33)

$$\epsilon = -\frac{1}{\pi} \int_{x_a}^{x_b} |k(x)|\, dx$$

(8.34)

for $E < V_0$. The coefficient, for a wave moving from left to right is, R_+ to the right and L_+ to the left of the barrier, respectively. The coefficient for a wave moving in the opposite direction is R_- and L_- respectively.

8.4. Derive the Wigner tunneling factor using the transmission probability for the parabolic barrier

$$P(E) = \frac{1}{1 + \exp(2\pi(V_0 - E)/\hbar\omega^{\#})} \tag{8.35}$$

i.e., evaluate

$$k(T) = \int_0^{\infty} d(\beta E) \exp(-\beta E) P(E) \tag{8.36}$$

where $\beta = 1/kT$. Use that

$$\int_{-\infty}^{\infty} \frac{\exp(-\mu x)}{1 + \exp(-x)} = \pi \csc(\mu\pi) \tag{8.37}$$

Discuss the assumptions made in arriving at the Wigner tunneling factor.

8.2   THE TRANSMISSION FACTOR

At equilibrium, the rate constant for flow over a dividing line between reactants and products is given by

$$k_e = \frac{1}{Q} \int_{p_a > 0} \frac{p_a}{m} \exp(-H(\{p_j\}, \{q_j\})/kT)\delta(s - s_0)\Pi \, dp_j \, \Pi \, dq_j \tag{8.38}$$

where Q is the reactant partition function per unit volume. The rate constant can now be obtained through

$$k = k_e \langle \kappa \rangle \tag{8.39}$$

where $\langle \kappa \rangle$ is an average transmission factor, which can be obtained by calculating the probability that trajectories starting at the dividing line s_0 will react, i.e.,

$$\langle \kappa \rangle = \frac{N_R}{N_t} \qquad (8.40)$$

where N_R is the number of trajectories N_t starting at the dividing line that react. For the six trajectories shown in Fig. 8.3, we see that two out of six are successful trajectories. Trajectories 2 and 3 are nonreactive; trajectory 6 is reactive, but not from left to right; and trajectory 4 is reactive from left to right, but should not be counted since, with a different choice of initial conditions at the dividing line, one would obtain trajectory 5, i.e., trajectory 5 would be counted twice. The variables at the dividing line should be taken from thermal distributions, and furthermore p_a should be greater than zero. Consider, for example, a linear saddle point complex A ... B ... A. We wish to start with this complex possessing energy in the symmetric (p) and asymmetric (p_a) stretch modes, noting that the asymmetric stretch will lead to reaction. The Hamiltonian is

$$H = \frac{p_a^2}{2m_a} + \frac{p^2}{2m} \qquad (8.41)$$

where $m_a = m_A(1 + 2m_A/m_B)/2$ and $m = m_A/2$. Thus p_a is [according to Eq.

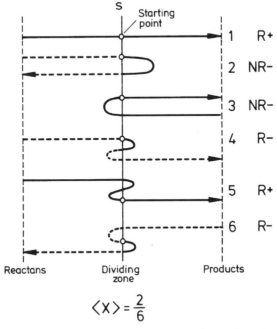

Fig. 8.3. The transmission factor is calculated by starting trajectories at the dividing line, and integrating forward and backwards in time.

(8.38)] taken from the normalized distribution

$$\frac{p_a}{m_a k T} \exp(-p_a^2/2m_a k T) \tag{8.42}$$

and p from $\exp(-p^2/2mkT)$. Thus we get

$$p_a = \sqrt{-2m_a k T \ln \xi} \tag{8.43}$$

$$p = \pm\sqrt{2mkTx} \tag{8.44}$$

where ξ is a random number between zero and one. The value of x is obtained from the equation $\text{erf}(x) = \xi$, where $\text{erf}(x)$ is the error function.

In order to obtain the exact classical reaction rate constant one has to evaluate the characteristic function $\chi(p, q)$ [63]. This function has the property of being unity if the trajectory is reactive, if it passes from reactants to products. Thus $\chi(p, q)$ is a function of the initial phase space variables (p, q). It may in practice, however, be determined by following trajectories backwards and forwards in time as described above. In this manner it is only necessary to integrate the dynamical equations over a much shorter time interval than would otherwise be necessary. In the transition state limit, the function $\chi(p, q)$ is replaced by the function $\theta(p_s)$, where $\theta(p_s) = 0$ for $p_s < 0$, and $\theta(p_s) = 1$ for $p_s > 0$.

EXERCISES

8.5. Derive Eq. (8.44) using that the error function is defined by

$$\text{erf}(x) = \frac{2}{\sqrt{\pi}} \int_0^x dt \, \exp(-t^2) \tag{8.45}$$

9

QUANTUM THEORY FOR RATE CONSTANTS

The correction to the transition state theory given in the previous chapter was introduced by evaluating the transmission factor. This was done by integrating trajectories started at the transition state backwards and forwards in time. In this manner the above barrier reflection could be accounted for. However, quantum mechanical effects such as tunneling were introduced approximately using semiclassical arguments and theories. Over the last 20 years Miller and coworkers [66, 67] have developed a quantum mechanical version of transition state theory, which allows calculation of reaction rate constants (which in principle are exact) without having to calculate the complete scattering matrix. The rate constant is instead calculated directly, by introducing quantum mechanical expressions for the flux over a dividing surface. In order to formulate the quantum mechanical version of the theory for obtaining reaction rate constants, it is convenient once again to consider the classical expression.

In classical theory, the thermal rate constant for a chemical reaction can be written as

$$k(T) = \frac{1}{Q(T)} \int d(\beta E) \, \exp(-\beta E) N(E, s_0) \qquad (9.1)$$

where $N(E, s_0)$ is the number of states passing a dividing line at $s = s_0$, along the reaction coordinate, per time unit, and where $\beta = 1/kT$. The number of states $N(E, s_0)$ can be computed by considering the integral

$$N(E, s_0) = \frac{1}{h^{3N-6}} \int ds \int dp_s \int d\{p_k\} \int d\{Q_k\} \delta(E - H) \delta(s - s_0) \frac{p_s}{m} \theta(p_s)$$

$$(9.2)$$

where $d\{p_k\}$ denotes a multiple differential $dp_1 \ldots dp_{3N-6}$, h is Planck's constant, m the mass connected to the motion along the reaction path s, p_s the momentum of this motion, and p_k and Q_k are the momenta and coordinates for the motion perpendicular to the reaction path motion. The above expression assumes a classical integration over phase space, subject to the constraint that the flux should go from reactants to products, i.e., the Heaviside function $\theta(p_s)$ is zero unless $p_s > 0$, where it is unity. In order to introduce a quantum mechanical description we note that the quantum "trace" is defined such that

$$Q(T) = \text{Tr}[\exp(-\beta H_0)] = \sum_n \exp(-\beta E_n) \frac{1}{h} \int_{-\infty}^{\infty} dp \, \exp(-\beta p^2/2\mu) \quad (9.3)$$

where H_0 is the Hamiltonian of the reactants A + BC, E_n the vibrational eigenstates of molecule BC, μ the reduced mass, and p the translational momentum (a collinear collision geometry is assumed). The integral over p arises from an approximation to the sum over translational states n_x, i.e.,

$$\sum_{n_x} \sim \frac{L_x}{h} \int dp \quad (9.4)$$

where we have used that $p = \hbar k_x$ and $k_x = 2\pi/L_x n_x$. Since we need the translational partition function per unit volume, we have, in Eq. (9.3), divided by L_x, the box length. Here we have considered a one-dimensional example, but Eq. (9.3) is readily extended to the three-dimensional case. The numerator in the above expression can be expressed as a quantum trace in a similar fashion

$$k(T) = \frac{1}{Q(T)} \text{Tr}[\exp(-\beta H) F \theta(p_s)] \quad (9.5)$$

where the flux operator is $F = \delta(s)p_s/m$. However, the quantization has not been properly introduced in this expression. The flux operator should have the symmetrized form given below in Eq. (9.17); and the function $\theta(p_s)$, which in the classical theory removes trajectories not proceeding from reactants to products in the proper way, should be replaced by a projection operator. This operator should project onto states that in the future ($t \to \infty$) have positive momenta, and thus move to the product side. Denoting the scattering state evolving from internal state n and translational state p as Ψ_{np}, the projection operator has the property [67]

$$P\Psi_{np} = \Psi_{np} \quad p > 0$$
$$P\Psi_{np} = 0 \quad\quad p < 0$$

Introducing the projection operator, we see that the rate constant can be written as

$$k(T) = \frac{1}{Q(T)} \sum_n \int_{-\infty}^{\infty} dp \, \langle \Psi_{np}|F \, \exp(-\beta H)P|\Psi_{np}\rangle \qquad (9.6)$$

The basis set Ψ_{np} is a complete set, in principle an arbitrary basis, but here taken to be the eigenbasis for the total Hamiltonian H. Hence

$$k(T) = \frac{1}{Q(T)} \, \text{Tr}[F \, \exp(-\beta H)P] \qquad (9.7)$$

where we note that the projection operator with the desired properties can be expressed in terms of the above basis as

$$P = \sum_n \int_{-\infty}^{\infty} dp \, \theta(p)|\Psi_{np}\rangle\langle\Psi_{np}| \qquad (9.8)$$

Since P is expressed in terms of the eigenbasis of H, we have $[H, P] = 0$, which allows us to write the expression for the rate constant in a more symmetric fashion

$$k(T) = \frac{1}{Q(T)} \, \text{Tr}[F \, \exp(-\beta H/2)P \, \exp(-\beta H/2)] \qquad (9.9)$$

The wave function $|\Psi_{np}\rangle$ evolves from a given reactant channel wave function in the past $|\Phi_{np}\rangle$ according to the Hamiltonian H for the system, i.e.,

$$\Psi_{np}(t) = \exp[-iH(t - t')/\hbar]\Phi_{np}(t') \qquad (9.10)$$

where $t \to \infty$. The wave function Φ_{np} evolves under the Hamiltonian H_0 such that

$$\Phi_{np}(t') = \exp[-iH_0(t' - t)/\hbar]\Phi_{np}(t) \qquad (9.11)$$

Combining the two equations and letting $t' = 0$ we have

$$|\Psi_{np}(t)\rangle = \lim_{t \to \infty} \exp(-iHt/\hbar)\ \exp(iH_0t/\hbar)|\Phi_{np}(t)\rangle \qquad (9.12)$$

We note that the initial wave functions $\Phi_{np}(t')$ all have momenta $p > 0$ (Fig. 9.1). This "property" will not be changed by propagation without interaction, i.e., with H_0. For the projection operator we then have

$$P = \exp(iHt/\hbar)\ \exp(-iH_0t/\hbar)P_0\ \exp(iH_0t/\hbar)\ \exp(-iHt/\hbar) \qquad (9.13)$$

where P_0 is the projection operator

$$P_0 = \sum_n \int_{-\infty}^{\infty} dp\ \theta(p)|\Phi_{np}\rangle\langle\Phi_{np}| \qquad (9.14)$$

in the time-propagated "initial" basis. Operating with P_0 is equivalent to operating with $\theta(p)$, since for the initial states $p > 0$, i.e., we can replace P_0 by $\theta(p)$. Furthermore since H_0 commutes with $\theta(p)$ we finally get

$$P = \lim_{t \to -\infty} \exp(iHt/\hbar)\theta(p)\ \exp(-iHt/\hbar) \qquad (9.15)$$

It is then convenient to replace $\theta(p)$ with $\theta(s)$. States which evolve with positive momentum to the product state will have $s > 0$ (for a formal proof, see [67]); but since s (the dividing line) is at this point arbitrary, we will always be able to find a value of s for which $\theta(s) = \theta(p)$ (e.g., if s is positioned well on the side of the reactants). Thus we note that the flux operator is, in principle, independent of where the dividing surface is placed. In practical calculations, the actual location will only affect how long the dynamics has to be followed in time. Replacing $\theta(p)$ with $\theta(s)$ we get

$$k(T) = \frac{1}{Q(T)} \lim_{t \to \infty} \text{Tr}\{F\ \exp[iH(t + i\hbar\beta/2)/\hbar]\theta(s)\ \exp[-iH(t - i\hbar\beta/2)\hbar]\} \qquad (9.16)$$

Using that $[H, \theta(s)] = \hbar/iF$, where F is the flux operator

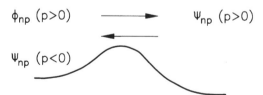

Fig. 9.1. The initial states Φ_{np} all have $p > 0$ (positive momenta). The scattered states are divided in positive and negative momentum states.

$$F = \frac{1}{2} \left(\frac{\hat{p}}{m} \delta(s) + \delta(s) \frac{\hat{p}}{m} \right) \qquad (9.17)$$

it is possible [67] to rewrite the flux–flux autocorrelation function $C_{\mathrm{ff}}(t)$, defined by

$$C_{\mathrm{ff}}(t) = \frac{d}{dt} \mathrm{Tr}\{F \; \exp[iH(t + i\hbar\beta/2)/\hbar]\theta(s) \; \exp[-iH(t - i\hbar\beta/2)/\hbar]\} \qquad (9.18)$$

in the form

$$C_{\mathrm{ff}}(t) = \mathrm{Tr}\left\{ F \; \exp\left[\left(-\frac{\beta}{2} + \frac{it}{\hbar} \right) H/\hbar \right] F \; \exp\left[\left(-\frac{\beta}{2} - \frac{it}{\hbar} \right) H/\hbar \right] \right\}$$

$$(9.19)$$

such that

$$k(T) = \frac{1}{Q(T)} \frac{1}{2} \int_{-\infty}^{\infty} C_{\mathrm{ff}}(t) \, dt \qquad (9.20)$$

The flux–flux autocorrelation function is a more symmetric expression than that of Eq. (9.16), which makes it easier to compute. Furthermore the rate constant is obtained as an integral over the flux–flux correlation function and hence the time behavior of $C_{\mathrm{ff}}(t)$ determines how long the dynamics have to be followed in time. Very often the correlation function decreases rapidly (e.g., exponentially) from its value at $t = 0$, and the dynamics have to be followed for only a short time in order to determine the rate constant.

Another symmetric expression is defined by

$$C_{ss}(t) = \mathrm{Tr}\{\theta(-s) \; \exp[iH(t + i\hbar\beta/2)/\hbar]\theta(s) \; \exp[-iH(t - i\hbar\beta/2)/\hbar]\} \qquad (9.21)$$

Using this expression we can write

$$k(T) = \frac{1}{Q} \lim_{t \to \infty} \frac{dC_{ss}(t)}{dt} \qquad (9.22)$$

The expression for the flux operator can more generally be given as [73]

$$\hat{F} = \frac{1}{2m} (\mathbf{n} \cdot \mathbf{p}\delta(f) + \delta(f)\mathbf{n} \cdot \mathbf{p}) \qquad (9.23)$$

where $\mathbf{n}$ is a unit vector perpendicular to a surface f in configuration space, which separates reactants and products, i.e.,

$$\mathbf{n} = \nabla f / |\nabla f| \tag{9.24}$$

Thus we are interested in calculating the flux over this surface in the long time limit (for $t \to \infty$ in Eq. (9.19)). In this way one is certain that all recrossing effects have been properly included. However, in practical calculations the time limit is, of course, finite and if grid methods (with finite boundaries) are used, then it is necessary to include absorbing complex potentials at the boundaries, in order to avoid nonphysical reflections [73].

9.1   THE PARABOLIC BARRIER PROBLEM

The expressions discussed above can in some cases be calculated analytically. As an example we mention the parabolic barrier problem with the following coordinate representation of the propagator

$$\langle s' | \exp(-iHt_c/\hbar) | s \rangle \tag{9.25}$$

where $t_c = t - i\hbar\beta/2$ and

$$H = -\frac{\hbar^2}{2m}\frac{d^2}{ds^2} + V_0 - \frac{1}{2}ks^2 \tag{9.26}$$

where $k = m\omega^2$. We note that Eq. (9.25) is the propagator for the harmonic oscillator with ω replaced by $i\omega$. Thus we can use the expression in Eq. (4.15) to get

$$\langle s' | \exp(-iHt_c/\hbar) | s \rangle = \sqrt{\frac{m\omega}{2\pi i\hbar \ \sinh(\omega t_c)}} \ \exp(-iV_0 t_c/\hbar)$$

$$\cdot \exp\left(\frac{im\omega}{2\hbar \ \sinh(\omega t_c)}\ [(s^2 + s'^2) \ \cosh(\omega t_c) - 2ss']\right) \tag{9.27}$$

One obtains

$$C_{ss}(t) = \int_0^\infty ds \int_{-\infty}^0 ds' \, |\langle s'| \exp(-iHt_c/\hbar)|s\rangle|^2$$

$$= \frac{kT}{h} \exp(-\beta V_0) \frac{\hbar\omega\beta/2}{\sin(\hbar\omega\beta/2)} f(t) \qquad (9.28)$$

where the Heaviside functions $\theta(-s)$ and $\theta(s)$ in Eq. (9.21) are zero, from 0 to ∞ and $-\infty$ to 0, respectively. The function $f(t)$ has the property that $d/dt f(t) \to 1$ for $t \to \infty$ (see the Exercise **9.1**). Thus the rate constant, obtained using Eq. (9.22), is

$$k(T) = \frac{1}{Q(T)} \exp(-\beta V_0) \frac{kT}{h} \frac{\hbar\omega\beta/2}{\sin(\hbar\omega\beta/2)} \qquad (9.29)$$

which is the correct result for the parabolic barrier (see Exercise 5.4). For a parabolic barrier, the flux–flux correlation function will decay towards zero as $\exp(-2\omega t)$ [67]. In general, of course, one has to resort to numerical evaluations of the flux–flux correlation function. Introducing for example a basis set such that $H|n\rangle = E_n|n\rangle$, we get

$$k(T) = \frac{1}{Q(T)} \lim_{t \to \infty} \hbar \sum_n \sum_{p \neq n} \frac{\sin[(E_p - E_n)t/\hbar]}{(E_p - E_n)}$$

$$\cdot \exp(\beta\hbar(E_p - E_n)/2)|\langle n|F|p\rangle|^2 \qquad (9.30)$$

In order to evaluate the infinite time limit while avoiding artificial reflections, it is necessary to introduce imaginary absorbing potentials at the edges of the interaction region (see e.g., ref. [73]).

EXERCISES

9.1. Show that Eq. (9.29) is valid for the parabolic barrier.

9.2. Show that for a free particle

$$C_{ff}(t) = \frac{kT}{h} \frac{(\hbar\beta/2)^2}{[t^2 + (\hbar\beta/2)^2]^{3/2}} \qquad (9.31)$$

9.2   RECOMBINATION REACTIONS

The flux–flux correlation method can also be formulated so as to account for recombination reactions [74], i.e.,

$$A + B + M \rightarrow AB + M \tag{9.32}$$

The mechanism for the recombination process is that the atoms A and B form a collision complex A $-$ B*, which is stabilized by a collision with the atom or molecule M:

$$AB* + M \rightarrow AB + M \tag{9.33}$$

with the rate constant $k*$. The recombination rate constant is obtained (classically) by the flux of A $-$ B* complexes passing a dividing surface, which separates the metastable and stable A $-$ B molecule. Thus

$$k(T) = \frac{1}{Q(T)} \int d(\beta E) \; \exp(-\beta E) N(E, s_0) P_{\text{sta}}(\omega, t^*) \tag{9.34}$$

where $N(E, s_0)$ is the number of states passing the dividing line s_0 per unit time, and $P_{\text{sta}}(\omega, t^*)$ is the probability that these will stabilize, i.e., suffer a stabilizing collision during the time interval from 0 to $t*$. This probability depends upon the collision frequency $\omega = [M]k*$ and, in the strong collision limit, where deactivation occurs in every collision, is given by

$$P_{\text{sta}}(\omega, t^*) = 1 - \exp(-\omega t^*) \tag{9.35}$$

which may be written as

$$P_{\text{sta}}(\omega, t^*) = \int_0^\infty dt \; \omega \; \exp(-\omega t)\theta(s, t) \tag{9.36}$$

where $\theta(s, t) = 1$ for $t < t^*$, and is zero otherwise. We now introduce an expression for the number of states passing the dividing line $s = s_0$ in the time interval 0 to t^*

$$N(E, s_0) = \frac{1}{h^{3N-6}} \int ds \int dp_s \int d\{p_k\} \int d\{Q_k\} \; \delta(E - H)\delta(s - s_0)(-p_s/m)$$
$$\cdot \theta(s, t) \tag{9.37}$$

and

$$\frac{d}{dt} \theta(s, t) = \dot{s} \frac{d}{ds} \theta(s, t) = \frac{p_s}{m} \delta(s - s(t)) \tag{9.38}$$

where we have used that the derivative of the Heaviside function is a delta function. We can then obtain, using partial integration,

$$k(T) = \frac{1}{Q(T)h^{3N-6}} \int_0^\infty dt \, \exp(-\omega t) \int ds \int dp_s$$

$$\cdot \int d\{Q_k\} \int d\{p_k\} \, \exp(-\beta H)\delta(s - s_0) \frac{p_s}{m} \delta(s - s(t)) \frac{p_s}{m} \quad (9.39)$$

Or quantum mechanically [74, 75]

$$k(T) = \frac{1}{Q(T)} \int_0^\infty dt \, \exp(-\omega t)C_{ff}(t) \quad (9.40)$$

where $C_{ff}(t)$ is the flux–flux correlation function as defined by Eq. (9.19). Thus the recombination rate is the Laplace transform (with respect to the collision frequency) of the flux–flux correlation function.

EXERCISE

9.3. Consider the following recombination scheme:

$$A + B \underset{k_i}{\overset{k_1}{\rightleftharpoons}} AB^*(i) \quad (9.41)$$

$$AB^*(i) + M \overset{k*}{\longrightarrow} AB + M \quad (9.42)$$

a. Express the recombination rate using the steady state assumption for [AB*]. Express the rate in terms of the rate constant k_i for unimolecular decay of AB* in state i and the collision frequency $\omega = k^*[M]$.
b. Find an expression for the flux–flux correlation function, which is consistent with the above rate constant.

10

STATISTICAL AND PHASE SPACE METHODS

The methods that have been previously discussed for obtaining the reaction rate constant for a chemical reaction all require knowledge either about the full multidimensional potential energy surface or parts of it, i.e., information on the transition state region or along the reaction path. For complex systems this information is either not available or extremely difficult to achieve. In such cases we need to prescribe avenues that require less information, e.g., because they involve assumptions such as the statistical assumption, where all parts of phase space are assumed to be equally probable. Thus we just have to calculate the available phase space for the reactants and the products; the reaction probability is taken to be the ratio between these two volumes. In a quantum mechanical formulation, the statistical assumption is that all quantum states of the collision complex are equally probable. However, the calculation of phase space and the number of quantum states is subject to some constraints—namely that total energy and angular momentum are conserved. Consider as an example the collision of an atom and a diatomic molecule

$$A + BC \rightarrow [A \ldots B \ldots C]^{\#} \rightarrow \text{products} \qquad (10.1)$$

The total cross section for the process α to β, where α and β denote reaction channels,

$$\sigma_{\alpha \rightarrow \beta} = \sum_{J} \sigma_{\alpha \rightarrow J} P_{J \rightarrow \beta} \qquad (10.2)$$

116

where the first factor $(\sigma_{\alpha \to J})$ is the cross section for forming a complex at a given value of the total angular momentum, and the last is the probability for breaking up to the channel β. Introducing the orbital angular momentum l and the rotational angular momentum of BC as j we get in the statistical theory

$$\sigma_{\alpha \to J} = \frac{\pi}{k_\alpha^2} \sum_l (2l+1) w_J(l,j) \tag{10.3}$$

where $\hbar^2 k_\alpha^2 / 2\mu = E_{\text{kin}}^\alpha$ is the initial kinetic energy, μ the reduced mass and $w_J(l,j)$ the ratio between the number of quantum states of the complex and that of the initial system, i.e.,

$$w_J(l,j) = \frac{2J+1}{(2l+1)(2j+1)} \tag{10.4}$$

The summation over orbital angular momentum is subject to the most restrictive of the two constraints, namely

$$|l - j| \leq J \leq j + l \tag{10.5}$$

$$l \leq l_{\max} \tag{10.6}$$

where $l_{\max}$ is determined through the effective potential (see Fig. 10.1). Note that $w_J = 1$ yields $\sigma = \pi b_{\max}^2$, where the maximum impact parameter $b_{\max}$ is related to $l_{\max}$ by $\sqrt{2\mu E_{\text{kin}}} b_{\max} = l_{\max}$. Introducing the statistical factor w_J given by Eq. (10.4) we get

$$\sigma_{\alpha \to J} = \frac{\pi}{k_\alpha^2} \frac{2J+1}{2j+1} \sum_{l^\alpha} 1 \tag{10.7}$$

where the sum counts the number of states in channel α that can form the complex with a given energy and total angular momentum J. Thus

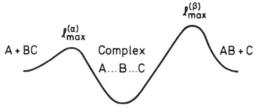

Fig. 10.1. The effective potential defines constraints for the summation over the number of states.

$$\sum_{j\alpha} 1 = N_{J,\alpha} = \sum_{n_\alpha} g_\alpha(n_\alpha; EJ) \tag{10.8}$$

where n_α is the vibrational quantum number of BC, and g_α the number of states for channel α obeying the constraints from Eqs. (10.5) and (10.6), (see Fig. 10.2). We note that l_{max}^α is (for a given n_α) a function of j^α through the kinetic energy

$$E_{kin}^\alpha = E - E_{n_\alpha}^{vib} - E_{j^\alpha}^{rot} + \Delta E_{ex}^\alpha \tag{10.9}$$

where the exothermicity ΔE_{ex}^α of the channel α is measured—for example, from the entrance channel. For a long-range potential of the type

$$V(R) = -C/R^n \tag{10.10}$$

where $n > 2$ we can estimate l_{max} from the effective potential

$$V_{eff} = \frac{\hbar^2 l(l+1)}{2\mu R^2} - \frac{C}{R^n} \tag{10.11}$$

by demanding that $E_{kin} > 0$ at R^*, defined by $(d V_{eff}/dR)_{R^*} = 0$. Thus we obtain

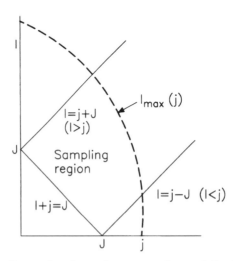

Fig. 10.2. The sampling region for a phase space theory defined by the constraints: $l \le l_{max}$ and $|l-j| \le J \le l+j$.

$$l_{max}(l_{max} + 1) = \frac{\mu}{\hbar^2} \left[\frac{E_{kin}^\alpha (nC)^{2/(n-2)}}{\frac{1}{2} - \frac{1}{n}} \right]^{(n-2)/n} \qquad (10.12)$$

We have introduced the number of states $N_{J,\alpha}$ for the entrance channel that can form a complex with a given energy and total angular momentum. Likewise, for channel β we can evaluate $N_{J,\beta}$. We now introduce the statistical assumption that the probability for breakup of the complex is proportional to the number of states that can be formed from a given channel, i.e.,

$$P_{J \to \alpha} = \text{const } N_{J,\alpha} \qquad (10.13)$$

The constant is

$$\text{const} = \frac{1}{\sum_{q=\alpha,\beta,\gamma} N_{J,q}} = \frac{1}{N_{tot}(J)} \qquad (10.14)$$

where α, β, and γ denote the three arrangement channels. Thus we get for the total cross section

$$\sigma_{\alpha \to \beta} = \sum_J \sigma_{\alpha \to J} \frac{N_{J\beta}}{N_{tot}(J)} = \frac{\pi}{k_\alpha^2 (2j+1)} \sum_J (2J+1) \frac{N_{J\alpha} N_{J\beta}}{N_{tot}(J)} \qquad (10.15)$$

This expression for the cross section has been obtained by Light [76]. A similar expression can be obtained using a statistical assumption on the S-matrix elements, where one introduces an average S-matrix $\langle S \rangle$ and the statistical assumption that

$$\langle |S_{\beta\alpha}|^2 \rangle = \frac{1}{N_J} \qquad (10.16)$$

where N_J is the number of possible final channels for a given angular momentum J [77].

10.1 PRIOR DISTRIBUTIONS AND SURPRISALS

The dynamical information which can be pulled out of a given set of experimental data depends upon the deviation from the statistical limit or the statistical prediction of the measured quantities. Thus the "prior" (or statistical) distribution is of importance, and the deviations from the statistical distribution functions are then used to analyze dynamical effects through so-called surprisal plots [see Eq. (10.29)]. For this reason it is of importance to be able to calculate the prior distribution functions. This is achieved by using the relation for the density $\rho(E_t)$ of translational states

$$\rho(E_t)\, dE_t = \frac{4\pi p^2}{h^3}\, dp \qquad (10.17)$$

where p is the translational momentum, and $\rho(E_t)\, dE_t$ is the number of states per unit volume in the energy range dE_t at the translational energy E_t. Consider a system consisting of two fragments a and b, comprised of N_a and N_b atoms, respectively. The prior distribution function for this system is, for a given energy E, the number of states for the system with a fixed internal rot/vib states, divided by the total number of states, i.e.,

$$P_0(n_k^a, n_k^b; E) = \frac{g(n_k^a, n_k^b)\sqrt{E - \epsilon(n_k^a, n_k^b)}}{\sum_{n_k^a, n_k^b} g(n_k^a, n_k^b)\sqrt{E - \epsilon(n_k^a, n_k^b)}} \qquad (10.18)$$

where $g(n_k^a, n_k^b)$ is the degeneracy factor associated with the quantum state n_k^a for fragment a, and n_k^b for fragment b. The term $E - \epsilon(n_k^a, n_k^b)$ is the translational energy of the relative motion. We have used the relation that, according to Eq. (10.17), $\rho \sim p$. It is possible to carry out the summation in the above equation numerically, but often it suffices to make a simple estimate using the rigid rotor and harmonic oscillator assumptions for the internal states. In this way we can obtain analytical expressions for the prior distributions. Consider, for example, an atom + a symmetric top molecule (with N atoms). The denominator (d) of the above equation is

$$d = \sum_{n_1 = 0} \cdots \sum_{n_M = 0} \sum_{J=0} \sum_{K=-J}^{J} (2J+1)\sqrt{E - \sum_{i=1}^{M} E_{n_i} - E_{\text{rot}}} \qquad (10.19)$$

where $M = 3N - 6$ for a nonlinear molecule, the degeneracy $g = 2J + 1$, and J and K are the rotational quantum numbers for a symmetric top molecule. The rotational and vibrational energies are now approximated as

$$E_{\text{rot}} = BJ(J+1) \sim B(J + 1/2)^2 \qquad (10.20)$$

$$E_{n_i} = \hbar \omega_i (n_i + 1/2) \qquad (10.21)$$

where B is a rotational constant. Let us introduce the fractional energies $f_R = E_{\text{rot}}/E$ and $f_i = E_{n_i}/E$, where E is the total energy. If, in addition, the summations are approximated by integrals we have

$$d \sim \frac{2E^{M+2}}{B^{3/2}\hbar^M \Pi_i \omega_i} \int_0^1 df_R \int_0^{1-f_R} df_1 \ldots \int_0^{1-f_R-\Sigma_{i=1}^{M-1} f_i} df_M$$

$$\sqrt{f_R\left(1 - f_R - \sum_{i=1}^{M} f_i\right)} \tag{10.22}$$

which can be evaluated (see Exercises 1 and 2) to give

$$d = \frac{E^{M+2}\pi}{2B^{3/2}\hbar^M (M+2)!\Pi_i \omega_i} \tag{10.23}$$

which then leads to the prior distribution function

$$P_0(f_R, f_1, \ldots, f_M) = \frac{4}{\pi}(M+2)!\sqrt{f_R\left(1 - f_R - \sum_i f_i\right)} \tag{10.24}$$

The distribution of one (or more) of the variables f_R, f_i can be obtained by integrating over the others. Thus for example we have

$$P_0(f_R) = \frac{4}{\pi}(M+2)!\sqrt{f_R} \int_0^1 df_1 \ldots \int_0^{1-f_R-\Sigma_{i=1}^{M-1} f_i} df_M$$

$$\sqrt{1 - f_R - \sum_{i=1}^{M} f_i} \tag{10.25}$$

which gives for the rotational energy distribution

$$P_0(f_R) = \frac{2}{\sqrt{\pi}} \frac{\sqrt{f_R}(M+2)!(1-f_R)}{\Gamma(M+3/2)} \tag{10.26}$$

where $\Gamma(x)$ is the gamma function. The prior (or the statistical) distributions can be used to obtain a first estimate of the probability of molecular processes.

The deviation from the prior distribution on a log-scale is the surprisal, which is the basis of an information theoretical approach to molecular collision theory. This approach has been developed by Levine and coworkers [78]. Thus the information content of a given distribution is given by

$$\Delta S = R \sum_{i=1}^{N} P_i \ln[P_i/P_i^0] \tag{10.27}$$

where R is the gas constant, P_i is the true probability (distribution), and P_i^0 the statistical one. Surprisal analysis considers the magnitude of

$$-\ln[P_i/P_i^0] = \lambda_0 + \sum_{r=1}^{n} \lambda_r A_r(i) \tag{10.28}$$

and expresses the deviation from the statistical result in terms of the n dynamical moments $A_r(i)$, where i denotes a specific event or outcome of the collision. The deviation of $\ln[P_i/P_i^0]$ from zero indicates that the process is nonstatistical, i.e., dynamical constraints are present. For inelastic events, one often encounters an exponential energy gap law. Here we would have

$$-\ln[P(i',i;T)/P_0(i',i;T)] = \lambda_0 + \lambda_1|E_i - E_{i'}|/kT \tag{10.29}$$

where i is the initial and i' the final state, and λ_0 and λ_1 are temperature-independent constants.

EXERCISES

10.1. Calculate the prior distribution for an atom plus a diatomic molecule. Express the prior distribution in terms of the fractional energy variables $f_i = E_i/E$, using the relation

$$\int_0^1 x^{a-1}(1-x)^{b-1}\,dx = \frac{\Gamma(a)\Gamma(b)}{\Gamma(a+b)} \tag{10.30}$$

10.2. Derive Eq. (10.23).

10.3. Show that Eq. (10.15) satisfies detailed balance, i.e, that $(2j_\alpha+1)k_\alpha^2\sigma_{\alpha\to\beta} = (2j_\beta+1)k_\beta^2\sigma_{\beta\to\alpha}$.

10.2   STOCHASTIC METHODS

Contrary to statistical theories, stochastic methods retain a dynamical concept. However, the dynamics is governed not by the Schrödinger or Hamiltons equations of motion, but by a more approximate equation—the Master equation or a Fokker-Planck equation. These equations deal with the dynamical evolution of probabilities rather than quantum amplitudes. Thus we can imagine the probability that a system will evolve from a set of quantum states $\mathbf{a}^0$ at time $t = 0$, to $\mathbf{a}$ at time t, with the probability $P(\mathbf{a}, \mathbf{a}_0, t)$. For the time evolution of the conditional probabilities it is possible to derive a Master equation [79]

$$\frac{\partial \rho(\mathbf{a}, \mathbf{a}_0, t)}{\partial t} = \sum_{\mathbf{a}'} A(\mathbf{a}, \mathbf{a}', t) P(\mathbf{a}', \mathbf{a}_0, t) \qquad (10.31)$$

The transition matrix $A(\mathbf{a}, \mathbf{a}', t)$ can formally be derived from the evolution operator

$$\Psi(t + \Delta t) = U(t + \Delta t, t) \Psi(t) \qquad (10.32)$$

Introducing a state expansion in a complete orthonormal basis ϕ_n we have

$$\Psi(t) = \sum_n a_n(t) \phi_n \qquad (10.33)$$

and hence

$$a_n(t + \Delta t) = \sum_m \langle \phi_n | U(t + \Delta t, t) | \phi_m \rangle a_m(t) \qquad (10.34)$$

For the probability $P_n(t) = |a_n(t)|^2$ we obtain

$$P_n(t + \Delta t) = \sum_m |\langle \phi_n | U(t + \Delta t, t) | \phi_m \rangle|^2 P_m(t) + \text{off-diagonal terms} \quad (10.35)$$

Considering the constraint $\sum_n P_n(t + \Delta t) = 1$, we see that it can be fulfilled if we assume that

$$\text{off-diagonal terms} = P_n(t) - \sum_m P_n |\langle \phi_m | U(t + \Delta t, t) | \phi_n \rangle|^2 \qquad (10.36)$$

which finally gives the following Master equation for the evolution of the probability distribution

$$\frac{P_n(t + \Delta t) - P_n(t)}{\Delta t} = \frac{1}{\Delta t} \sum_m |\langle \phi_n | U(t + \Delta t, t) | \phi_m \rangle|^2 (P_m - P_n) \qquad (10.37)$$

where we have used that the evolution operator is Hermitian. We are now able to introduce the transition matrix

$$A(n, m, t) = \lim_{\Delta t \to 0} \frac{1}{\Delta t} |\langle \phi_n | U(t + \Delta t, t) | \phi_m \rangle|^2 \qquad (10.38)$$

The transition matrix will be diagonally dominant, i.e., it can be expanded as

$$A(n, m, t) = A(n, n, t) + \left. \frac{\partial A(n, m, t)}{\partial m} \right|_{m=n} (m - n)$$

$$+ \frac{1}{2} \left. \frac{\partial^2 A(n, m, t)}{\partial m^2} \right|_{m=n} (m - n)^2 + \ldots \qquad (10.39)$$

Introducing this expansion allows us to obtain the Fokker-Planck equation

$$\frac{\partial P(n, t)}{\partial t} = B_0(n, t) + B_1(n, t) \frac{\partial P(n, t)}{\partial n} + B_2(n, t) \frac{\partial^2 P(n, t)}{\partial n^2} \qquad (10.40)$$

where the coefficients $B_k(n, t)$ are related to the elements of the transition matrix (see Exercise **10.4**). We shall below consider a simple case where the problem can be solved analytically. In more general cases, approximations must be introduced in order to evaluate the n- and t-dependence of the coefficients $B_k(n, t), k = 0, 1, 2$.

10.2.1   The Forced Oscillator

Consider as an example the forced oscillator treated previously (Chapter 2). The total wave function can be expanded as

$$\psi(r, t) = \sum_n a_n(t) \phi_n(r) \exp(-iE_n t/\hbar) \qquad (10.41)$$

where $\phi_n(r)$ is an oscillator eigenfunction and E_n the corresponding eigenvalue. Introducing the coupling $V(r, t)$, we obtain from the time-dependent Schrödinger equation the following set of coupled equations:

$$i\hbar \frac{da_n(t)}{dt} = \sum_m V_{nm}(t) \exp(i\epsilon_{mn}t) a_m(t) \qquad (10.42)$$

where $V_{nm}(t) = \langle n|V(r,t)|m\rangle$ and $\epsilon_{nm} = (E_n - E_m)/\hbar$. Considering that the probability $P_n = |a_n(t)|^2$ we have

$$\frac{d}{dt} P_n(t) = \frac{2}{\hbar^2} \sum_m \int dt'\, F_{nm}(t,t') P_m(t) + \sum_m G_{nm}(t,t') \qquad (10.43)$$

where

$$F_{nm}(t,t') = V_{nm}(t) V_{nm}(t') \cos(\epsilon_{nm}(t - t')) \qquad (10.44)$$

and G_{nm} are off-diagonal terms. We have also made use of the assumption that the correlation between $V_{nm}(t)$ and $V_{nm}(t')$ is short compared to the timescale over which the probabilities (or the amplitudes) change, i.e., we assume that

$$P_m(t) \sim a_m^*(t) a_m(t') \sim a_m(t) a_m^*(t') \qquad (10.45)$$

The last term in Eq. (10.43) represents all the off-diagonal terms, which can be determined by noting that conservation of probability yields

$$0 = \sum_n \dot{P}_n \qquad (10.46)$$

Imposing this constraint we obtain

$$G_{nm}(t,t') = -\frac{1}{\hbar^2} P_n(t) \int dt'\, F_{nm}(t,t') \qquad (10.47)$$

and substituting in the above equation we finally get

$$\dot{P}_n(t) = \frac{1}{\hbar^2} \sum_m \int_{-\infty}^{t} dt'\, F_{nm}(t,t') (P_m(t) - P_n(t)) \qquad (10.48)$$

This Master equation can be further approximated by a Fokker-Planck equation. This is done by expanding the kernel and the probabilities in a Taylor series in the quantum numbers [79]. This can also be illustrated in the above example by introducing the following expressions, valid for the linearly forced oscillator

$$V_{nm} = f(t)\sqrt{2m}\,\delta_{m,n\pm1} \tag{10.49}$$

We then get

$$\dot{P}_n(t) = D(t)[nP_{n-1} + (n+1)P_{n+1} - (2n+1)P_n] \tag{10.50}$$

where the diffusion coefficient $D(t)$ is given by

$$D(t) = \frac{4}{\hbar^2} \int_{-\infty}^{t} dt'\, f(t)f(t')\cos(\omega(t-t')) \tag{10.51}$$

Introducing a continuous variable n we get the Fokker-Planck equation

$$\frac{\partial P(n,t)}{\partial t} = D(t)\,\frac{\partial}{\partial n}\left[\left(n+\frac{1}{2}\right)\frac{\partial P(n,t)}{\partial n}\right] \tag{10.52}$$

The solution to this equation can be obtained analytically (see Exercise **10.5**).

EXERCISES

10.4. Determine the coefficients $B_k(n,t)$ from the matrix elements $A(n,m,t)$.

10.5. Show that the Eq. (10.52) has the solution

$$P(n,t) = \frac{1}{\eta(t)}\exp\left[-\frac{1}{\eta(t)}(n+n_0+1)\right]$$

$$\cdot I_0\left(\frac{2}{\eta(t)}\sqrt{\left(n+\frac{1}{2}\right)\left(n_0+\frac{1}{2}\right)}\right), \tag{10.53}$$

where n_0 denotes the initial state $P(n,-\infty) = \delta_{n,n_0}$, and $\eta(t) = \int_{-\infty}^{t} D(t')\,dt'$. Find an expression in the limit of small values of $\eta(t)$. Note that the Bessel function $I_0(z) \sim \exp(z)$ for large values of the argument.

11

PHOTODISSOCIATION

Photodissociation processes are often treated within the Born-Oppenheimer approximation (see Chapter 6), the electric dipole approximation for the interaction between light and the system of electrons and nuclei. Usually the (weak field) approximation for this interaction is first order. Thus, assuming that the transition from the electronic state 1 to 2 is induced by a field varying with time as

$$\mathbf{d} \cdot \mathbf{E}_0 \cos(\omega t) \tag{11.1}$$

where $\nu = \omega/2\pi$ is the frequency of the light, $\mathbf{E}_0$ the electric field strength, and $\mathbf{d}$ the electric dipole operator, and within the weak field limit, we obtain the following first order expression (see Section 2.1) for the probability for inducing a transition from state 1 to 2:

$$P_{12} = |\alpha^+|^2 = \frac{1}{\hbar^2} \left| \int_0^t dt' \, \exp(i\omega_{21}t) \, \langle 2|\mathbf{d} \cdot \mathbf{E}_0|1\rangle \cos \omega t \right|^2 \tag{11.2}$$

where $\omega_{21} = (E_2 - E_1)/\hbar$. Evaluating the time integral we get

$$P_{12} \sim \frac{|\langle 2|\mathbf{d} \cdot \mathbf{E}_0|1\rangle|^2}{\hbar^2} \frac{\sin^2[(\omega_{21} - \omega)t/2]}{(\omega_{21} - \omega)^2} \tag{11.3}$$

where the rapidly oscillating terms containing $\omega_{21} + \omega$ have been neglected,

127

since they contribute insignificantly to the absorption probability. Introducing now that the Dirac delta function can be represented as

$$\delta(x) = \frac{2}{\pi} \lim_{t \to \infty} \frac{\sin^2(xt/2)}{x^2 t} \tag{11.4}$$

we can write

$$P_{12} \sim \frac{|\langle 2|\mathbf{d} \cdot \mathbf{E}_0|1\rangle|^2}{\hbar^2} \frac{\pi}{2} t\delta(\omega_{12} - \omega) \tag{11.5}$$

The transition rate obtained is

$$k_{12} = \frac{dP_{12}}{dt} = \frac{|\langle 2|\mathbf{d} \cdot \mathbf{E}_0|1\rangle|^2}{\hbar^2} \frac{\pi}{2} \delta(\omega - \omega_{12}) \tag{11.5a}$$

We are now able to obtain the energy absorbed per unit volume and per unit time in a cavity of volume V, and with number of absorbers N. It is given by

$$\frac{dW}{dt} = -\frac{N}{V} \hbar\omega_{21} k_{12} \tag{11.6}$$

where W is the energy density of the field. In the absence of magnetic fields, W is [80]

$$W = \frac{\epsilon}{2} E_0^2 \tag{11.7}$$

where ϵ is the permittivity. Let us introduce the intensity as $I = Wc$ and $dW/dt = dI/(c\,dt) = dI/dz$ and the Beer-Lambert absorption law

$$\frac{dI}{dz} = -\rho\sigma(\omega)I(z) \tag{11.8}$$

for a light beam traveling along the z-axis where ρ is the density of absorbers N/V. This defines the absorption cross section as

$$\sigma(\omega) = \frac{\pi}{\hbar} \frac{\omega}{\epsilon c} \delta(\omega - \omega_{12})|\langle 2|\mathbf{e} \cdot \mathbf{d}|1\rangle|^2 \tag{11.9}$$

where the integration is over electronic and nuclear coordinates, c is the speed of light, and $\mathbf{e}$ a unit vector in the direction of the electric field, i.e., $\mathbf{E}_0 = \mathbf{e}E_0$.

Introducing further the dipole moment function

$$\mu_{21}(\mathbf{X}) = \int d\mathbf{x} \, \phi_2^*(x, \mathbf{X}) \, \mathbf{d} \, \phi_1(\mathbf{x}, \mathbf{X}) \tag{11.10}$$

where $\phi_{1,2}$ denote two Born-Oppenheimer electronic states, we can write

$$\langle 2|\mathbf{e} \cdot \mathbf{d}|1 \rangle = \langle mE|\phi_i \rangle \tag{11.11}$$

where the last brackets denote integration over the nuclear coordinates only, and where we have introduced the "initial" state

$$|\phi_i\rangle = \mathbf{e} \cdot \mu_{21}|E_i\rangle \tag{11.12}$$

where $|E_i\rangle$ denotes the ground state nuclear wave function with energy E_i.

Thus the partial photodissociation cross section is given by a Franck-Condon expression (see for example ref. [69])

$$\sigma_m(\omega) = C\omega |\langle m, E|\phi_i \rangle|^2 \tag{11.13}$$

where C is a constant (including the δ-function) and $|E, m\rangle$ a scattering state on the upper electronic surface (see Fig. 11.1), i.e.,

$$\hat{H}|E, m\rangle = E|E, m\rangle \tag{11.14}$$

where

$$\hat{H} = \hat{T}_{\text{kin}} + \Delta_{12} + V_2 + D_e \tag{11.15}$$

$\hat{T}_{\text{kin}}$ is the kinetic energy operator and V_2 the potential energy on the upper surface (i.e., $\lim_{r \to \infty} V_2(r) = 0$). The term Δ_{12} is the asymptotic energy separation between surfaces 1 and 2, and D_e is the "dissociation" energy on the lower surface (see Fig. 11.1). The transition dipole moment is the electronic dipole moment for transition between the two electronic states 1 and 2; it depends on the internuclear coordinates but is often approximated by a constant.

The total energy is

$$E = E_i + \hbar\omega \tag{11.16}$$

where ω is the frequency of the absorbed light. The total absorption cross section is obtained by summing the partial ones

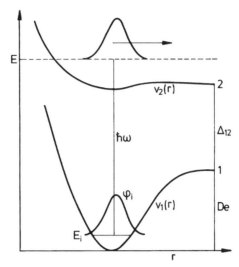

Fig. 11.1. The ground state wavefunction is promoted to the upper surface and propagated in time.

$$\sigma(\omega) = \sum_m \sigma_m(\omega) \tag{11.17}$$

It is possible and often convenient to express the absorption cross section in terms of a time-dependent theory. Thus by using

$$1 = \int dE \, \delta(\hbar\omega + E_i - E) = \int dE \, \frac{1}{2\pi\hbar} \int dt \, \exp(i\omega t + iE_i t/\hbar - iEt/\hbar) \tag{11.18}$$

and the closure relation

$$1 = \int dE \sum_m |E,m\rangle\langle m, E| \tag{11.19}$$

it is possible to write

$$\sum_m |Em\rangle\langle mE| = \frac{1}{2\pi\hbar} \int dt \, \exp(i\omega t + iE_i t/\hbar - iHt/\hbar) \tag{11.20}$$

where in the exponential we have replaced E by the operator H. This is necessary, since the left-hand side is an operator. From Eqs. (11.13) and (11.17) one

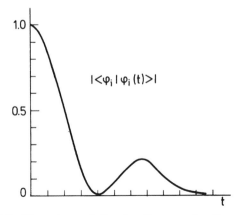

Fig. 11.2. The autocorrelation function as a function of time.

then obtains

$$\sigma(\omega) = C'\omega \int_{-\infty}^{\infty} dt \, \exp(i\omega t + iE_i t/\hbar)\langle \phi_i| \exp(-i\hat{H}t/\hbar)|\phi_i\rangle \qquad (11.21)$$

Thus we see that the total absorption cross section can be obtained as

$$\sigma(\omega) = C'\omega \int_{-\infty}^{\infty} dt \, \exp(i\omega t + iE_i t/\hbar)\langle \phi_i|\phi_i(t)\rangle \qquad (11.22)$$

where $\phi_i(t)$ is the time-propagated initial wave function, propagated on the upper electronic state with the Hamiltonian $\hat{H}$. This propagation can be carried out using the time-dependent propagators mentioned previously (Chapter 5). We see that the total absorption cross section is obtained as the Fourier transform of the time correlation function of the propagated and the initial wave function. A typical time evolution of the correlation function is shown in Fig. 11.2.

The propagation on the upper surface is started at $t = 0$ and carried on until the correlation function vanishes. In order to obtain the Fourier transform in the time interval from minus to plus infinity we use the relation $\langle \phi_i|\phi_i(-t)\rangle = \langle \phi_i|\phi_i(t)\rangle^*$.

11.1 STRONG-FIELD LIMIT

The interaction between radiation and matter can, within the electric dipole approximation, be expressed in terms of the dipole moment operator

$$\mathbf{d} = -e \sum_i \mathbf{x}_i + \sum_n Z_n \mathbf{X}_n \tag{11.23}$$

where $\mathbf{x}_i$ and $\mathbf{X}_n$ denote electronic and nuclear coordinates, respectively, and Z_n the nuclear charge. The interaction H_{int} is obtained as [70]

$$H_{int} = \mathbf{d} \cdot \prod \tag{11.24}$$

where

$$\prod = -ic \sum_{\mathbf{k},\lambda} k \mathbf{e}^{(\lambda)}(\mathbf{k})[b_{k\lambda} \exp(i\mathbf{k} \cdot \mathbf{r}) - b_{k\lambda}^\dagger \exp(-i\mathbf{k} \cdot \mathbf{r})] \tag{11.25}$$

where $\mathbf{e}^{(\lambda)}(\mathbf{k})$ are field polarization vectors, $\mathbf{k}$ the wave vector, $b_{k\lambda}$ the field quantization operators (labeled by the photon mode index k and λ the polarization direction), and c the speed of light. The total wave function can now be expanded as

$$\Psi(\mathbf{x}, \mathbf{X}, t) = \sum_{in} \xi_{in}(\mathbf{X}, t)\phi_i(\mathbf{x}; \mathbf{X})|n_{k\lambda}\rangle \tag{11.26}$$

where ϕ_i are electronic adiabatic states and $|n_{k\lambda}\rangle$ a photon state. This expansion is inserted in the time-dependent Schrödinger equation (TDSE) using the expression for the total Hamiltonian

$$H = H_0 + H_{int} + H_{rad} + T_N \tag{11.27}$$

where H_0 is the Born-Oppenheimer Hamiltonian, i.e.,

$$H_0\phi_i(\mathbf{x}, \mathbf{X}) = E_i(\mathbf{X})\phi_i(\mathbf{x}, \mathbf{X}) \tag{11.28}$$

The radiation field Hamiltonian H_{rad} can be expressed in terms of the field operators

$$H_{rad} = \sum_{k\lambda} \hbar\omega_{k\lambda} b_{k\lambda}^\dagger b_{k\lambda} \tag{11.29}$$

and the term T_N is the nuclear kinetic energy. From the TDSE one obtains a set of coupled equations in the expansion coefficients ξ_{in}, but in order to simplify the discussion it is convenient to introduce a classical treatment of the field, i.e., to assume that the electric field vector in the vicinity of **r** can be approximated as

$$E(\mathbf{r}, t) = \mathbf{E}_0 A(t) \, \cos(\omega t) \tag{11.30}$$

where $A(t)$ is a "pulse" shape function. The interaction term in the Hamiltonian then becomes

$$H_{\text{int}} = -\mathbf{d} \cdot \mathbf{E} \tag{11.31}$$

The expansion is now simplified to

$$\Psi(\mathbf{x}, \mathbf{X}, t) = \sum_i \xi_i(\mathbf{X}, t) \, \phi_i(\mathbf{x}, \mathbf{X}) \tag{11.32}$$

Assuming just two electronic states we obtain from the TDSE

$$i\hbar \, \frac{\partial \xi_1(\mathbf{X}, t)}{\partial t} = (T_N + E_1(\mathbf{X}))\xi_1 - \mathbf{d}_{11} \cdot \mathbf{E} \, \xi_1 - \mathbf{d}_{12} \cdot \mathbf{E} \, \xi_2 \tag{11.33}$$

$$i\hbar \, \frac{\partial \xi_2(\mathbf{X}, t)}{\partial t} = (T_N + E_2(\mathbf{X}))\xi_2 - \mathbf{d}_{22} \cdot \mathbf{E} \, \xi_2 - \mathbf{d}_{21} \cdot \mathbf{E} \, \xi_1 \tag{11.34}$$

where we have integrated over electronic coordinates and neglected nonadiabatic coupling terms and where

$$\mathbf{d}_{ij}(\mathbf{X}) = \int d\mathbf{x} \, \phi_i^*(\mathbf{x}, \mathbf{X}) \, \mathbf{d} \, \phi_j(\mathbf{x}, \mathbf{X}) \tag{11.35}$$

is the dipole moment function.

We now assume that the system initially is in state I on the lower surface, such that

$$\Psi(\mathbf{X}, \mathbf{x}, t) = \phi_1(\mathbf{x}, \mathbf{X})\xi_{1I}(\mathbf{X}) \, \exp\left(-\frac{i}{\hbar} \, E_{1I} t\right) \tag{11.36}$$

In the strong-field limit, we would solve the coupled equations [Eqs. (11.33–11.34)] while for weak fields we may introduce a first order treatment, inserting the wave function $\xi_1(\mathbf{X}, t) = \xi_{1I} \, \exp(-iE_{1I}t/\hbar)$ where

$$[T_N + E_1(\mathbf{X})]\xi_{1I}(\mathbf{X}) = E_{1I}\xi_{1I}(\mathbf{X}) \tag{11.37}$$

in Eq. (11.34) to obtain

$$i\hbar\xi_2(\mathbf{X}, t) = -E_0 \int_0^t dt' A(t') \cos(\omega t') \exp\left(-\frac{i}{\hbar} E_{1l}t\right) \exp\left(-\frac{i}{\hbar} H_2 t'\right)$$
$$\cdot \, \mathbf{d}_{12} \cdot \mathbf{e} \, \xi_{1l}(\mathbf{X}) \tag{11.38}$$

From the solution ξ_2 we obtain the total or partial absorption cross sections by projection on the appropriate asymptotic eigenfunctions.

11.1.1   Asymptotic Projection

The shape and structure of photoabsorption spectra give information on the potential energy surface of the "upper" electronic state on which the dissociation process occurs. More information is available if the photoabsorption cross section can be resolved in the vibrational/rotational states of the products, i.e., if partial cross sections can be measured.

Considering for example the photodissociation of a triatomic molecule

$$ABC + \hbar\omega \rightarrow A + BC \tag{11.39}$$

at an energy $E = E_i + \hbar\omega$, the scattering states of the fragment channels are

$$\frac{1}{rR} Y_{lm_l}(\Theta \Phi) \sqrt{\frac{\mu}{\hbar k_{vj}}} \exp(-ik_{vj}R)\phi_{vj}(r)Y_{jm_j}(\theta\phi) \tag{11.40}$$

for $R \rightarrow \infty$. Here

$$\frac{\hbar^2 k_{vj}^2}{2\mu} = E - E_{vj} \tag{11.41}$$

and $\mu = m_A(m_B + m_C)/(m_A + m_B + m_C)$. The term $\phi_{vj}(r)$ is the vibrational eigenfunction of the diatomic fragment. The orientation of the vectors $\mathbf{r}$ and $\mathbf{R}$ are in a space-fixed coordinate system given by the polar angles θ, ϕ and Θ, Φ, respectively. The eigenfunctions connected to the rotational and orbital motions are Y_{jm_j}, and Y_{lm_l}, respectively. Instead of using these asymptotic states it is convenient to introduce states labeled by J, the total angular momentum, and its projection on a space-fixed axis M. This can be achieved through vector coupling of $\mathbf{j}$ and $\mathbf{l}$, i.e., introducing [72]

$$Y_{jl}^{JM} = \sqrt{2J + 1}(-1)^{j-l-M} \sum_{m_j m_l} Y_{jm_j}(\theta, \phi) Y_{lm_l}(\Theta, \Phi) \begin{pmatrix} j & l & J \\ m_j & m_l & -M \end{pmatrix}$$

$$(11.42)$$

where $\begin{pmatrix} \vdots \end{pmatrix}$ is a 3-j symbol.

Introducing now the spherical harmonics $Y_{j\mu}(\eta, \xi)$, where the angles η and ξ specify the orientation of the diatomic molecule in a coordinate system with the z-axis along $\mathbf{R}$ (i.e., a body-fixed frame), we have [72]

$$Y_{jm_j}(\theta, \phi) = \sum_{\mu} (-1)^{m_j - \mu} Y_{j\mu}(\eta, \xi) D^j_{-m_j, -\mu}(\Theta, \Phi) \qquad (11.43)$$

where $D^j_{m_j\mu}(\Theta, \Phi)$ are matrix elements of a rotation operator, which rotates the basis functions $Y_{jm_j}(\theta, \phi)$ from a space-fixed to the body-fixed frame [72]. Introducing further

$$Y_{lm_l}(\Theta, \Phi) = \sqrt{(2l + 1)/4\pi} D^l_{-m_l 0}(\Theta, \Phi) \qquad (11.44)$$

we finally get [72]

$$Y_{jl}^{JM} = (-1)^{j+l-M} \sqrt{\frac{(2J + 1)(2l + 1)}{4\pi}} \sum_{\mu} \begin{pmatrix} j & l & J \\ \mu & 0 & -\mu \end{pmatrix} Y_{j\mu}(\eta, \xi) D^J_{-M, \mu}(\Theta, \Phi)$$

$$(11.45)$$

These are the Arthur-Dalgarno states [71]. Thus the asymptotic wave function may be expanded to

$$\Psi \to \frac{1}{Rr} \sum_{vjl} Y_{jl}^{JM}(\eta\xi; \Theta\Phi) u^J_{vjl}(R)\phi_{vj}(r) \qquad (11.46)$$

where the radial functions $u^J_{vjl}(R)$ are plane waves for $R \to \infty$. For the special case of $J = M = 0$ the asymptotic states simplify to

$$\frac{1}{rR} \sum_{vj} Y_{j0}(\eta, 0)\phi_{vj}(r)u^J_{vj0}(R) \qquad (11.47)$$

where we have used the relation [72]

$$\begin{pmatrix} j & l & 0 \\ 0 & 0 & 0 \end{pmatrix} = \delta_{jl} \frac{(-1)^j}{\sqrt{2j+1}} \tag{11.48}$$

and incorporated the factors $(-1)^j$ and $1/\sqrt{4\pi}$ in the expansion coefficients.

EXERCISE

11.1. Show that Eq. (11.4) is a proper definition of a δ function.

11.2 THE RRGM METHOD

The absorption spectrum can, as we have seen in the previous section, be calculated by time-propagation on the excited state potential energy surface of the wave function multiplied with the transition dipole moment. This propagation can be carried out in the "standard fashion" by using one of the time propagators mentioned previously. However, there is an alternative—namely the RRGM (recursive residue generation method). Introducing the eigenstates of the total Hamiltonian, i.e.,

$$H\psi_\alpha = E_\alpha \psi_\alpha \tag{11.49}$$

allows us to expand the wave function $\phi(t)$ propagated on the upper surface as

$$|\phi(t)\rangle = \sum_\alpha c_\alpha(t)|\psi_\alpha\rangle \tag{11.50}$$

where

$$c_\alpha(t) = c_\alpha(t=0)\ \exp(-iE_\alpha t/\hbar) \tag{11.51}$$

This follows from the fact that $|\psi_\alpha\rangle$ are eigenfunctions of H. We furthermore have

$$c_\alpha(0) = \langle\psi_\alpha|\phi(t=0)\rangle \tag{11.52}$$

where $|\phi(t=0)\rangle = |\phi_0\rangle$ is the initial wave function, i.e.,

$$|\phi_0\rangle = \sum_\alpha \langle\psi_\alpha|\phi_0\rangle|\psi_\alpha\rangle \tag{11.53}$$

Introducing $|\psi(t)\rangle$ and $|\phi_0\rangle$ in the expression for the photoabsorption cross section Eq. (11.22) we obtain

$$\sigma(\omega) = C\omega \int dt \, \exp(i\omega t) \langle \phi_0 | \phi(t) \rangle$$

$$= C\omega \int dt \, \exp(i\omega t) \sum_\alpha |\langle \psi_\alpha | \phi_0 \rangle|^2 \exp(-iE_\alpha t/\hbar) \qquad (11.54)$$

Thus we see that in order to evaluate the absorption cross section we need only to evaluate the residues $\langle \alpha | \phi_0 \rangle$, where we have used the notation $|\alpha\rangle$ instead of $|\psi_\alpha\rangle$. In the RRGM method one uses the assumptions that

(1) The number of initial states to be considered is small (perhaps only a single one).
(2) The coupling matrix is sparse.
(3) The residues $\langle \alpha | \phi_0 \rangle$ can be calculated (see below) from either two sets of eigenvalues or the first element of the eigenvectors of the tridiagonal matrix obtained from a Lanczos iteration.

Considering now Green's operator $1/(z - H)$ we have the matrix element

$$G_u(z) = \left\langle u_0 \left| \frac{1}{z - H} \right| u_0 \right\rangle = \sum_\alpha \frac{1}{z - E_\alpha} \langle u_0 | \alpha \rangle^2 \qquad (11.55)$$

and we see that the resolvent R_u is

$$R_u = \langle u_0 | \alpha \rangle^2 = \lim_{z \to E_\alpha} (z - E_\alpha) G_u(z) \qquad (11.56)$$

We now introduce the basis function of H where u_0 is the first (the input vector) in a series generated by a Lanczos iteration, i.e.,

$$|u_0\rangle, |u_1\rangle \cdots |u_M\rangle \qquad (11.57)$$

where M is the number of Lanczos iterations (see Chapter 5) and $M \ll N$, N being the number of states. Thus we can, in this basis, set up a matrix equation for the components of the Green's function

$$(z\mathbf{I} - \mathbf{H})\mathbf{G}(z) = \mathbf{I} \qquad (11.58)$$

where we need to evaluate $G_{00} = \langle u_0 | (z - H)^{-1} | u_0 \rangle = G_u(z)$, the first element of the matrix [see Eq. (11.56)]. This element can be obtained as the ratio of the determinants of two matrices

$$G_u(z) = \frac{\det[z - H]_r}{\det[z - H]} \tag{11.59}$$

where the index r denotes that the matrix is reduced by removing the first row and the first column. Since the determinant is given by its eigenvalues we have

$$G_{00} = \frac{(z - \epsilon_1^{(r)}) \cdots (z - \epsilon_{N-1}^{(r)})}{(z - \epsilon_1) \cdots (z - \epsilon_N)} \tag{11.60}$$

We see that all that is required is evaluating the eigenvalues of two matrices. The first matrix has the first column and the first row removed. The second matrix is the full matrix. Since the evaluation of the eigenvalues is an N^2 process as far as the computational effort is concerned, considerable reduction in computing time (a factor of N) is achieved as compared with the diagonalization of the H-matrix. For large systems spurious eigenvalues can occur when using the Lanczos reduction technique—but methods do exist for handling these problems [81].

Instead of using the above scheme, which requires the determination of the eigenvalues of **H**, another possibility exists. This method is based upon a diagonalization of the tridiagonal matrix obtained from the Lanczos recursion. The first vector in the recursion is $|u_0\rangle = |\phi_0\rangle$, and in order to evaluate the absorption cross sections we need to propagate the initial vector under the Hamiltonian H, i.e., we must compute

$$\langle u_0 | \exp(-iHt) | u_0 \rangle \tag{11.61}$$

The Lanczos recursion vectors tridiagonalize the Hamiltonian, i.e.,

$$\mathbf{L^+ H L} = \mathbf{T} \tag{11.62}$$

where **T** is a tridiagonal matrix and **L** the matrix containing the recursion vectors. Since these vectors by construction are orthogonal, i.e.,

$$\mathbf{L L^+} = \mathbf{I} \tag{11.63}$$

where **I** is the unit matrix, we can write

$$\begin{aligned} &\langle u_0 | \mathbf{L L^+} \exp(-i\mathbf{H}t) \mathbf{L L^+} | u_0 \rangle \\ &= \langle u_0 | \mathbf{L} \exp(-i\mathbf{T}t) \mathbf{L^+} | u_0 \rangle \end{aligned} \tag{11.64}$$

Since by construction (by the Lanczos iteration) u_0 is the first column vector in $\mathbf{L}$ and is orthogonal to the remaining vectors, we have

$$\langle u_0 \mathbf{L} | = \langle 10 \ \ldots \ 0 | \tag{11.65}$$

and hence

$$\langle u_0 | \exp(-iHt) | u_0 \rangle = \exp(-iT_{00}t) \tag{11.66}$$

i.e., only a single element of the T-matrix is needed.

The tridiagonal matrix $\mathbf{T}$ can be diagonalized by the orthogonal transformation with the matrix $\mathbf{A}$ such that

$$\mathbf{A^{+}TA} = \mathbf{D} \tag{11.67}$$

where $\mathbf{D}$ is a diagonal matrix. The eigenbasis to H is then column vectors of the matrix $\mathbf{LA}$. Thus the residue is obtained as

$$\langle u_0 | \beta \rangle = \sum_i u_0(i)\beta(i) = \sum_i L_{i1} \sum_{k=1}^{M} L_{ik} A_{k\beta} = A_{1\beta} \tag{11.68}$$

where $|\beta\rangle$ denotes an eigenvector to $\mathbf{H}$, i denotes the ith component of the vector, and we have used the orthogonality of the Lanczos recursion vectors, i.e.,

$$\sum_i L_{i1} L_{ik} = \delta_{1k} \tag{11.69}$$

Thus we only need the first element of each eigenvector of the tridiagonal matrix $\mathbf{T}$. If the complete set of vectors are needed, the numerical effort required to find the eigenvalues and the first component of the eigenvectors is an M^2 process rather than an M^3 operation.

For the photodissociation problem the survival residue $\langle n|\alpha\rangle\langle\alpha|n\rangle$ is required. Other time-dependent problems require the transition residues of the type

$$\langle n|\alpha\rangle\langle\alpha|n'\rangle \tag{11.70}$$

where $n \neq n'$. Often only a few of the transitions are important for a given problem and they can be obtained by using residue algebra [81], i.e.,

$$\langle n|\alpha\rangle\langle\alpha|n'\rangle = \tfrac{1}{2} \left[R_u(\alpha, n, n') - R_v(\alpha, n, n') \right] \tag{11.71}$$

where

$$R_u(\alpha, n, n') = \langle u|\alpha\rangle^2, \tag{11.72}$$

$|u\rangle = \frac{1}{\sqrt{2}}(|n\rangle + |n'\rangle)$ and $|v\rangle = \frac{1}{\sqrt{2}}(|n\rangle - |n'\rangle)$. Thus we need to start the Lanczos recursion with two different initial vectors, $|u\rangle$ and $|v\rangle$.

EXERCISE

11.2. Derive Eq. (11.59) by considering a simple two-level system, where

$$\begin{bmatrix} z - H_{00} & H_{01} \\ H_{10} & z - H_{11} \end{bmatrix} \begin{bmatrix} G_{00} & G_{01} \\ G_{10} & G_{11} \end{bmatrix} = \begin{bmatrix} 1 & 0 \\ 0 & 1 \end{bmatrix} \tag{11.73}$$

12

DENSITY OPERATORS

This chapter gives a brief introduction to the use of density operators within physical chemistry, focusing on the description of an evolving molecular system coupled to an outer medium. Density operators are a necessity when investigating the randomness of the real world; they form a basic building block within quantum statistical mechanics.

The density operator for a system having the state vector $|\psi(t)\rangle$ is defined by the outer product

$$\rho(t) = |\psi(t)\rangle\langle\psi(t)| \tag{12.1}$$

The mean of an operator A is then given as

$$\langle A \rangle = \langle\psi(t)|A|\psi(t)\rangle = \mathrm{Tr}\{A\rho(t)\} \tag{12.2}$$

The measurable information in the state vector is clearly contained in the density operator. The importance of the density operator approach is clearly seen for a case where the state of the system is not a pure state, a pure state having a specific state vector such as $|\psi(t)\rangle$.

The randomness of the real world manifests itself when a system is prepared with probabilities $P(i)$ for being in the different states $|\psi_i\rangle$. The expectation value of a quantum mechanical operator A is given by

$$\langle A \rangle = \Sigma_i P(i)\langle\psi_i|A|\psi_i\rangle \tag{12.3}$$

and defining a density operator as

141

$$\rho = \Sigma_i P(i) |\psi_i\rangle \langle \psi_i| \tag{12.4}$$

leads to

$$\langle A \rangle = \mathrm{Tr}\{A\rho\} \tag{12.5}$$

Thus the density operator contains both the statistical and quantum mechanical information about the system.

12.1   PROPERTIES OF DENSITY OPERATORS

In this section we consider some general properties of the density operator, the first being

$$\mathrm{Tr}\{\rho\} = 1 \tag{12.6}$$

since

$$\mathrm{Tr}\{\rho\} = \Sigma_i P(i) \langle \psi_i | \psi_i \rangle = \Sigma_i P(i) = 1 \tag{12.7}$$

Furthermore for an arbitrary state $|K\rangle$

$$\langle K|\rho|K\rangle = \Sigma_i P(i) |\langle K|\psi_i\rangle|^2 \geq 0 \tag{12.8}$$

which tells us that ρ is semidefinite.

For the situation where the ensemble of states $|\psi_i\rangle$ contains only one state $|S\rangle$ the density operator is given by

$$\rho = |S\rangle\langle S| \tag{12.9}$$

which, as mentioned previously, is said to be a pure state. In this case the system is then always in this state $|S\rangle$ and we have that

$$\rho^2 = |S\rangle\langle S||S\rangle\langle S| = \rho \tag{12.10}$$

The final property is that

$$\mathrm{Tr}\{\rho^2\} \leq 1 \tag{12.11}$$

where the equal sign holds only for a pure state.

Next we establish an equation of motion for the density operator starting from the time derivative of Eq. (12.4)

$$\frac{d}{dt}\,\rho(t) = \frac{d}{dt}\,(\Sigma_i\,P(i)\,|\psi_i(t)\rangle\langle\psi_i(t)|) \tag{12.12}$$

and utilizing the time-dependent Schrödinger equation for the state vector $|\psi_i(t)\rangle$

$$i\hbar\,\frac{d}{dt}\,|\psi_i(t)\rangle = H|\psi_i(t)\rangle \tag{12.13}$$

where the Hamiltonian of the system is given by H. We obtain that

$$\frac{d}{dt}\,\rho(t) = -\frac{1}{i\hbar}\,[H,\rho(t)] \tag{12.14}$$

which leads to the formal solution

$$\rho(t) = \exp\left(\frac{iH}{\hbar}\,t\right)\,\rho(0)\,\exp\left(-\frac{iH}{\hbar}\,t\right) \tag{12.15}$$

EXERCISES

12.1. Given that

$$\rho^2 = \rho \tag{12.16}$$

prove that we have a pure state.

12.2. Prove the validity of the final property of the density operator

$$\mathrm{Tr}\{\rho^2\} \leq 1 \tag{12.17}$$

where the equal sign applies to the pure state.

12.3. Derive Eq. (12.14).

12.2 QUANTUM STATISTICAL MECHANICS

The concept of a density operator for describing the quantum mechanical states of a system not in a pure state is only one aspect of quantum statistical mechanics. We need, in addition, to introduce the concepts of temperature, entropy, and statistical ensembles. This will be done through the maximization of entropy.

A macrostate with a very large number of particles is given by appropriate macroscopic variables: temperature, pressure, volume, and momentum. The microscopic picture of a macrostate consists of an enormous number of possible microscopic states. The microscopic states are specified by the energy, momentum, etc., of the individual particles. The concept of entropy gives a measure of how many different microstates lead to the same macrostate.

Let us consider a quantum mechanical system where the density operator ρ is written as

$$\rho = \begin{pmatrix} p_1 & 0 & 0 & 0 & \cdots \\ 0 & p_2 & 0 & 0 & \cdots \\ 0 & 0 & p_3 & 0 & \cdots \\ 0 & 0 & 0 & p_4 & \cdots \\ \vdots & \vdots & \vdots & \vdots & \end{pmatrix} \qquad (12.18)$$

and where it has been assumed that it is possible to find a basis where the density operator is diagonal.

Let us assume that we have M different quantum states that contribute equally to the density operator, which leads to a definition of entropy, S

$$S = k \, \log M \qquad (12.19)$$

where k is Boltzmann's constant. Also given that

$$p_i = \frac{1}{M} \qquad \text{for i} \leq M$$

$$p_i = 0 \qquad \text{for i} \geq M \qquad (12.20)$$

we have the following

$$\mathrm{Tr}\{\rho \, \log \rho\} = \Sigma_i \, p_i \, \log p_i = -\log M \qquad (12.21)$$

The trace is invariant with respect to unitary transformation of the basis and for this case [Eq. (12.20)] we deduce the expression for the entropy S

$$S = -k\mathrm{Tr}\{\rho \, \log \rho\} \qquad (12.22)$$

Clearly the probability distribution indicated by Eq. (12.20) is not always valid and this fact keeps one from knowing how many quantum states contribute to the density matrix. On the other hand Eq. (12.22) gives an indication of the average number of states contributing to the density matrix.

The fundamental concept of obtaining thermodynamic equilibrium through maximization of the entropy has been covered and discussed in many different textbooks [84–88]. We will make use of this concept in illustrating how to define partition functions within quantum statistical mechanics. Thermodynamic equilibrium of a system is achieved through a stationary state of the system. This tells us that the density operator is time-independent, and for a Hamiltonian without explicit time dependence, the Hamiltonian and the density operator commute with each other.

In order to proceed, we make the assumption that we are working in a representation that makes the density operator diagonal, thus for which

$$\rho = \Sigma_i \, p_i |i\rangle \langle i| \tag{12.23}$$

$$H = \Sigma_i \, E_i |i\rangle \langle i| \tag{12.24}$$

$$S = -k \, \Sigma_i \, p_i \, \log p_i \tag{12.25}$$

The constraints that must be fulfilled for the maximization of the entropy are those of mean energy content

$$\langle E \rangle = \Sigma_i \, p_i \, E_i |i\rangle \langle i| \tag{12.26}$$

and the normalization of the density operator

$$\text{Tr}\{\rho\} = \Sigma_i \, p_i = 1 \tag{12.27}$$

The maximization of the entropy subject to the two constraints leads to the Lagrangian

$$\delta(\Sigma_i \, p_i \, \log p_i + \alpha \, \Sigma_i \, p_i + \beta \, \Sigma_i \, E_i \, p_i) = 0 \tag{12.28}$$

having the solution

$$p_i = \exp(-\alpha - \beta E_i) \tag{12.29}$$

which leads to an expression for the density operator

$$\rho = \exp(-\alpha - \beta H) \tag{12.30}$$

We determine the two multipliers through the normalization constraint, Eq. (12.27) and the definition

$$\beta = \frac{1}{kT} \tag{12.31}$$

where β is related to the temperature T through the constant k, Boltzmann's constant. We then obtain the canonical density operator

$$\rho = \frac{1}{Z(\beta)} \, \exp(-\beta H) \tag{12.32}$$

where

$$Z(\beta) = \text{Tr}\{\exp(-\beta H)\} \tag{12.33}$$

which is the canonical partition function.

Another relevant ensemble is the grand canonical ensemble, which is obtained using a similar procedure, though an additional constraint is maintained: the particle number operator N is assumed to be a constant of motion, when maximizing the entropy. This gives the grand canonical density operator

$$\rho = \frac{1}{Z(\beta, \mu)} \exp[-\beta(H - \mu N)] \tag{12.34}$$

where μ is the chemical potential and the grand canonical partition function is given as

$$Z(\beta, \mu) = \text{Tr}\{\exp[-\beta(H - \mu N)]\} \tag{12.35}$$

EXERCISE

12.4. Prove Eq. (12.21).

12.3   SUBSYSTEMS WITHIN A SYSTEM

We are often concerned with a rather large total system consisting of a finite number of degrees of freedom of interest to us, as well as an enormous number of degrees of freedom of little importance to us. This is a common situation within statistical mechanics, where one often is concerned with a small system coupled to a heat bath kept at a given temperature. Chemical reactions in solution exemplify this quite nicely: the reactive molecules, constituting the system, are surrounded by the non-reacting part of the solvent, constituting the heat bath. The heat bath is so large that the system has limited influence on it. Yet, exchange of energy between the heat bath and the system takes place leading to dissipation and fluctuations.

Formally we will consider the situation in which we have a clear separation between the system and the heat bath and let

$$|s, b\rangle = |s\rangle|b\rangle \tag{12.36}$$

where s and b represent the system and bath degrees of freedom, respectively. Furthermore, we have operators acting separately either on the system or on the bath. The operators of the former are of course the ones of primary interest, since we are investigating the system and the bath is the environment within which the system undergoes changes. The mean of an operator, A, for the system with respect to an arbitrary state $|X\rangle$ is

$$\langle X|A|X\rangle = \Sigma_{b,b'} E_{s,s'} \langle X|s,b\rangle \langle s,b|A|s',b'\rangle \langle s',b'|X\rangle$$
$$= \Sigma_b \Sigma_{s,s'} \langle X|s,b\rangle \langle s|A|s'\rangle \langle s',b|X\rangle \qquad (12.37)$$

using the fact that

$$\langle s,b|A|s',b'\rangle = \langle s|A|s'\rangle \delta_{b,b'} \qquad (12.38)$$

where $\delta_{b,b'}$ is the Kronecker delta-function.

We express $\langle X|A|X\rangle$ as

$$\langle X|A|X\rangle = \text{Tr}_{\text{sys}} \{\rho_{\text{sys}} A\} \qquad (12.39)$$

where

$$\rho_{\text{sys}} = \Sigma_b \Sigma_{s,s'} |s'\rangle \langle s',b|X\rangle \langle X|s,b\rangle \langle s| \qquad (12.40)$$

which is the reduced density operator for the system and is formally written as

$$\rho_{\text{sys}} = \text{Tr}_b \{|X\rangle \langle X|\} \qquad (12.41)$$

The trace is taken only with respect to the bath degrees of freedom. We are able to rewrite this expression through the system states $|X_b\rangle$

$$|X_b\rangle = \sum_s \frac{|s\rangle \langle s,b|X\rangle}{\sqrt{Q(b)}} \qquad (12.42)$$

where

$$Q(b) = \Sigma_s |\langle s,b|X\rangle|^2 \qquad (12.43)$$

which gives the following expression for the reduced density operator

$$\rho = \Sigma_b Q(b)|X_b\rangle \langle X_b| \qquad (12.44)$$

Examining Eq. (12.44) we note that the reduced density operator has the same properties as those outlined in Section 12.1 concerning properties of density operators.

If the total system is described by a density operator ρ then the reduced density is given as

$$\rho_{\text{sys}} = \text{Tr}_b \{\rho\} \qquad (12.45)$$

13

EVOLUTION OF A TOTAL SYSTEM

We reconsider the situation where the total system contains two subsystems: the system under investigation and the heat bath. The total system could be a chemical reaction taking place in solution, where the reacting molecules constitute the system and the surrounding inert solvent the heat bath. The Hamiltonian of the total system is of the form

$$H = H_{sys} + H_{bath} + H_{int} \tag{13.1}$$

where H_{sys} is the Hamiltonian for the system, H_{bath} is the Hamiltonian for the heat bath, and H_{int} describes interactions between the two subsystems. The total density operator, ρ_{tot} for the system and the bath fulfills the following equation of motion

$$\frac{d\rho_{tot}(t)}{dt} = -\frac{i}{\hbar}\,[H, \rho_{tot}(t)] = -\frac{i}{\hbar}\,[H_{sys} + H_{bath} + H_{int}, \rho_{tot}(t)] \tag{13.2}$$

In order to determine expectation values of properties of the investigated system we wish to establish an equation of motion for the reduced density operator

$$\rho_{red}(t) = \text{Tr}_b\{\rho_{tot}(t)\} \tag{13.3}$$

where we perform the trace with respect to the degrees of freedom of the bath. Initially we transform Eq. (13.2) to the interaction picture using $H_{sys} + H_{bath}$

$$\rho_{int}(t) = \exp\left[\frac{i}{\hbar}(H_{sys} + H_{bath})t\right]\rho_{tot}(t)\exp\left[-\frac{i}{\hbar}(H_{sys} + H_{bath})t\right] \quad (13.4)$$

which can be reduced to

$$\frac{d\rho_{int}(t)}{dt} = -\frac{i}{\hbar}[H_{int}(t), \rho_{int}(t)] \quad (13.5)$$

where

$$H_{int}(t) = \exp\left[\frac{i}{\hbar}(H_{sys} + H_{bath})t\right]H_{int}\exp\left[-\frac{i}{\hbar}(H_{sys} + H_{bath})t\right] \quad (13.6)$$

This gives the following expression for the reduced density operator

$$\rho_{red}(t) = \text{Tr}_b\left\{\exp\left[-\frac{i}{\hbar}(H_{sys} + H_{bath})t\right]\rho_{int}(t)\exp\left[\frac{i}{\hbar}(H_{sys} + H_{bath})t\right]\right\}$$

$$(13.7)$$

The bath Hamiltonian H_{bath} depends only on the degrees of freedom of the heat bath. By using the cyclic property of the trace we find

$$\rho_{red}(t) = \exp\left[-\frac{i}{\hbar}H_{sys}t\right]\rho^*(t)\exp\left[\frac{i}{\hbar}H_{sys}t\right] \quad (13.8)$$

where

$$\rho^*(t) = \text{Tr}_b\{\rho_{int}(t)\} \quad (13.9)$$

which is the reduced density operator in the interaction picture.

We assume that the total system is prepared in such a manner that the initial conditions for the systems can be written as

$$\rho_{tot}(0) = \rho_{sys}(0)\rho_{bath}(0) \quad (13.10)$$

thereby assuming that the system and bath initially are independent, the density operator $\rho_{bath}(t)$ describes the heat bath. We also assume that the heat bath is sufficiently large so that its statistical properties are unaffected by the presence of the investigated system. This assumption is based on the weak coupling between the system and the heat bath and leads to

$$\rho_{tot}(0) = \rho_{sys}(0) \otimes \rho_{bath} \tag{13.11}$$

where ρ_{bath} is time-independent and $\otimes$ indicates a direct product.

In order to proceed, we return to Eq. (13.5) and integrate it twice

$$\rho_{int}(t) = \rho_{int}(0) - \frac{i}{\hbar} \int_0^t dt' [H_{int}(t'), \rho_{int}(0)]$$

$$- \frac{1}{\hbar^2} \int_0^t dt' \int_0^{t'} dt'' [H_{int}(t'), [H_{int}(t''), \rho_{int}(t'')]] \tag{13.12}$$

It appears that it could be interesting to iterate in this manner and generate a power series with respect to the interaction H_{int}; however, this could quickly become a cumbersome task. Instead we will differentiate Eq. (13.12) to give

$$\frac{d\rho_{int}(t)}{dt} = -\frac{i}{\hbar} [H_{int}(t), \rho_{int}(0)] - \frac{1}{\hbar^2} \int_0^t dt' [H_{int}(t), [H_{int}(t'), \rho_{int}(t')]]$$

$$\tag{13.13}$$

which is an integro-differential equation that will make it easy for us to obtain simple expressions for the reduced density operator.

Next we perform the trace with respect to the bath degrees of freedom to get

$$\frac{d\rho^*(t)}{dt} = -\frac{1}{\hbar^2} \int_0^t dt' \, \mathrm{Tr}_b \{[H_{int}(t), [H_{int}(t'), \rho_{int}(t')]]\} \tag{13.14}$$

where we have used Eqs. (13.9) and (13.10) and assumed that

$$\mathrm{Tr}_b \{H_{int}(t) \rho_{int}(0)\} = 0 \tag{13.15}$$

which means that the interaction operator does not contain diagonal elements in the representation that makes H_{bath} diagonal. In the case of diagonal elements within H_{int}, one can always include these in H_{sys} and thereby fulfill Eq. (13.15).

As stated previously we assume that the interactions between the heat bath and the system are sufficiently small so the density operator for the heat bath is not changed by the interactions. We will then assume, based on Eq. (13.10), that

$$\rho_{int}(t') \approx \rho^*(t') \otimes \rho_{bath} \tag{13.16}$$

The underlying assumptions are that the bath density operator is not significantly changed by interactions between the bath and the system, yet that the system density operator is allowed to change significantly since the system is much smaller than the bath, therefore, the effects of the interactions have a much larger impact on the system than on the bath. It is, however, possible to go beyond the assumption concerning the direct product approximation of the density operator; this will be covered in the chapter on electron transfer (Chapter 17).

Let us introduce the approximation from Eq. (13.16) in Eq. (13.14), which gives

$$\frac{d\rho^*(t)}{dt} = -\frac{1}{\hbar^2} \int_0^t dt' \, \mathrm{Tr}_b\{[H_{\mathrm{int}}(t), [H_{\mathrm{int}}(t'), \rho^*(t') \otimes \rho_{\mathrm{bath}}]]\} \qquad (13.17)$$

We assume H_{int} to be written as an operator sum of terms like $S_{\mathrm{sys}}T_{\mathrm{bath}}$. The operators S_{sys} and T_{bath} work only on the system and bath, respectively. This means that we will have terms like

$$\mathrm{Tr}_b\{[T_{\mathrm{bath}}(t), [T_{\mathrm{bath}}(t'), \rho_{\mathrm{int}}(t')]]\} \approx \rho(t') \otimes \mathrm{Tr}_b\{[T_{\mathrm{bath}}(t), [T_{\mathrm{bath}}(t'), \rho_{\mathrm{bath}}]]\}$$
$$(13.18)$$

which illustrates that bath correlation functions are not significantly changed by the presence of H_{int}. By making use of the assumption of the weak interaction we let the rate of change of the system density operator in the interaction picture be slow compared to changes in the bath operators $T_{\mathrm{bath}}(t), T_{\mathrm{bath}}(t')$. In the interaction picture the bath operators vary according to the bath Hamiltonian H_{bath}, and the bath correlation times will generally be determined by the choice of ρ_{bath}, such as the ensemble type and the temperature of the heat bath. The correlation times are in this case much shorter than the timescale that involves changes in the system density operator and in this case we can say that the factor $\rho(t')$ changes insignificantly during the decay period of the correlation functions represented by terms of the following character $T_{\mathrm{bath}}(t)T_{\mathrm{bath}}(t')$. Letting

$$\rho(t') \rightarrow \rho(t) \qquad (13.19)$$

amounts to the Markov approximation [89]. Utilizing it we are then able to write Eq. (13.17) as

$$\frac{d\rho^*(t)}{dt} = -\frac{1}{\hbar^2} \int_0^t dt' \, \mathrm{Tr}_b\{[H_{\mathrm{int}}(t), [H_{\mathrm{int}}(t'), \rho^*(t) \otimes \rho_{\mathrm{bath}}]]\} \qquad (13.20)$$

and we also let t be much larger than the thermal correlation times, thereby letting the lower limit of the integral go to $-\infty$ while using $\tau = t - t'$, which gives

$$\frac{d\rho^*(t)}{dt} = -\frac{1}{\hbar^2} \int_0^\infty d\tau \, \mathrm{Tr}_b \{[H_{\mathrm{int}}(t), [H_{\mathrm{int}}(t - \tau), \rho^*(t) \otimes \rho_{\mathrm{bath}}]]\}. \quad (13.21)$$

The Markov approximation tells us that knowledge of $\rho(t)$ at $t = t_o$ and use of Eq. (13.21) enable us to determine $\rho(t)$ for any $t > t_o$. In order to use Eq. (13.21) it is necessary to know the form of the Hamiltonian $H_{\mathrm{sys}}, H_{\mathrm{bath}}, H_{\mathrm{int}}$. When using this approach for investigating the evolution of a physical system it is important to bear in mind the short bath-correlation times and the use of perturbation theory.

14

NONEQUILIBRIUM SOLVATION

Within the last decade theoretical investigations have reached a level where the computed results successfully have reproduced, interpreted, and predicted many gas phase experiments. This level of competence should be the goal for the theoretical methods involving calculations of molecular properties of solvated molecules and chemical reactions in solutions as well, given that the great bulk of measurements are carried out for condensed phase samples, either in solvated, crystallized, or polymerized forms. Theoretical methods have to take account of how a solvated molecule, the solute, is affected by its environment, the solvent. The solvent interaction can be both long- and short-range in nature and the solvent-induced changes can change the vacuum property even in a qualitative way.

Initially, theoretical investigations of solvent effects considered solvation energies of molecular charge distributions [90–93]. These analytical models were extended to include electronic excitation and deexcitation processes [94–97]. Later, the methods developed into numerical approaches for solving the quantum mechanical equations involving the molecular system and the solvent [98–110]. In general, many-electron descriptions of the solute and the outer solvent have evolved around three model types:

(1) Supermolecular models
(2) Continuum models
(3) Semicontinuum models

Supermolecular models involve electronic structure calculations of a molecule solvated by a cluster of solvent molecules. The problems associated with supermolecular models are basis set superposition errors, the neglect of long-range effects, the choice of cluster structure, and identification of the effects—the latter point being related to the problems of identifying whether the effects originate from the solvent cluster or from the solute. As an example, one could consider the problem of calculating the hyperpolarizabilities or magnetizabilities of a solute in a supermolecular system. Both properties refer to the response of the total supermolecular system and not just the response of the solute towards the externally applied field.

Continuum models are based on the notion that the solvated molecule induces polarization charges in an outer medium. The induced polarization gives rise to an extra potential at the position of the solvated molecule. The reaction field from the induced polarization charges interacts with the solute and affects the molecular properties. The continuum models are generalizations of Kirkwood's original model [91], where point charges are enclosed by a spherical cavity and embedded in a structureless polarizable medium described by the macroscopic bulk dielectric constant. Generalizations include (1) a multipole expansion of the molecular charge distribution within the cavity from which the reaction field is generated [102, 108–112], and (2) a self-consistent quantum mechanical treatment of the reaction field interaction with the cavity molecules. The methods are usually termed the self-consistent reaction field model, SCRF, or the multiconfiguration self-consistent reaction field model, MCSCRF. These models neglect the short-range interactions between solvent molecules and solute, and one encounters problems with choice of the actual form of the cavity.

Semicontinuum models are appropriate for taking account of local and specific solvent effects around the solute [113–116]. The basic idea is to combine the two other approaches by surrounding the solute with a cluster of solvent molecules and embedding this supermolecule in a dielectric medium. This method enables both a description of the long-range polarization effects and of the short-range effects (e.g., hydrogen bonding, and charge transfer and exchange).

For the continuum and semicontinuum models the dielectric medium surrounding the molecular system is characterized by two dielectric constants ϵ_{op} and ϵ_{st}, accounting for the optical and total polarization, respectively.

Recently, rigorous quantum formulations have been established, and in these both the solute electronic modes and a continuum set of solvent optical polarization vectors are treated as fully quantum objects [117, 118]. The methods encompass the description of a large class of solvent mediated solute processes, e.g., solvatochromatic shifts of photon absorption and emission in a solute. In modeling these processes one encounters a broad range of relative relaxation times which describes how fast the different modes of the solvent are able to respond and participate in the evolution of the process [117–122].

From a general electrostatic free-energy functional [123–126] for the diluted solute–solvent system and a coherent state representation for the solute polarization field [117] we form a standard first order variational procedure. The variation is performed with respect to the total solute wave function and the optical polarization eigenvectors, and the final result is a nonlinear Schrödinger equation for the electronic state of the solute interacting with the field of a reaction potential and an integral equation for the optical contribution of the polarization eigenvectors.

The model gives two equilibrated quantized fields: one due to the solute electrostatic charge distribution and another due to the solvent response to this charge distribution through the optical polarization vector. A third field is present, the inertial polarization, which is slow to incorporate the changes in the electronic structure of the composite solute–solvent system.

Within the general theory of Kim and Hynes [117] one finds two limiting cases related to the ratio

$$\frac{\text{Intrinsic transition times within the quantized solvent}}{\text{Intrinsic transition times within the electronic state of the solute}} \quad (14.1)$$

For chemical or physical processes where the solvent optical polarization field fluctuates with a frequency higher than a characteristic transition frequency of an active electron of the solute, one is in the limit where each electronic state of the solute is equilibrated simultaneously with a complete set of solvent optical polarization eigenvectors. The complete set of solvent optical polarization eigenvectors only sees a slowly varying charge distribution of the solute.

In the other limiting case, the electronic state of the solute interacts with only one single eigenvector of the quantized optical polarization field. This is because the solvent optical polarization field possesses a longer fluctuation time compared to the timescale of changes in the electronic configuration of the solute. The solvent optical polarization eigenvector is the one that has been equilibrated with a given reference (ground or excited state). In this case one recovers the semiclassical (SC) nonequilibrium or equilibrium approaches for the polarization continuum model of the solvation effect [117, 127–138].

Within the framework of quantum chemistry an efficient polarizable continuum solvent–solute interaction model has been developed within both semiempirical and *ab initio* frameworks, using the SC approximation in its simplest equilibrium version [139]. Current implementations in quantum chemical programs are either dipole fields, extended multipole expansion fields, or fully numeric (point charge) electrostatic fields; and the encapsulating cavities are either spherical, ellipsoidal [130, 133–137, 140–143], or deformed cavities [131, 132, 144, 145]. Recently, Cossi et al. [146] developed a nonhomogeneous solvent model by allowing the dielectric constant to be a function of the modulus of the radial coordinate.

Of late, nonequilibrium solvation models have been established and implemented both for Hartree-Fock or multiconfigurational *ab initio* solute electronic wave functions. The present chapter outlines the theoretical background for these approaches.

14.1   THE MULTICONFIGURATIONAL SELF-CONSISTENT REACTION FIELD METHOD

14.1.1   The Homogeneous Dielectric Continuum Semiclassical Solvent Model

Within the SC approximation, we consider the solute and optional solvation shells enclosed in a cavity immersed in a dielectric medium with two types of polarizations, optical and inertial [138]. The electronic structure of the molecular complex within the cavity is represented by two multiconfigurational self-consistent field (MCSCRF) electronic wave functions, $\Psi_i(\mathbf{r})$ and $\Psi_f(\mathbf{r})$—these being the electronic wave functions for the initial and final states, respectively. The initial state of the system corresponds to a situation where the molecular charge distribution and the two polarization fields are in equilibrium with each other. Therefore

$$\mathbf{P}_{op}(\mathbf{r}) \equiv \mathbf{P}_{op}^{eq}[\mathbf{r}; \rho_i(\mathbf{r})] \tag{14.2}$$

$$\mathbf{P}_{in}(\mathbf{r}) \equiv \mathbf{P}_{in}^{eq}[\mathbf{r}; \rho_i(\mathbf{r})] \tag{14.3}$$

where $\rho_i(\mathbf{r})$ is the charge distribution of the molecular complex in the initial state corresponding to the electronic wave function $\Psi_i(\mathbf{r})$, and $\mathbf{P}_{op}(\mathbf{r})$, and $\mathbf{P}_{in}(\mathbf{r})$ are the equilibrated optical and inertial polarization vectors for the initial state. The total polarization vector, $\mathbf{P}(\mathbf{r})$ is given as

$$\mathbf{P}(\mathbf{r}) = \mathbf{P}_{op}(\mathbf{r}) + \mathbf{P}_{in}(\mathbf{r}) \tag{14.4}$$

The general electrostatic free-energy functional [125] $\mathcal{F}[(\mathbf{P}_{op}(\mathbf{r}), \mathbf{P}_{in}(\mathbf{r}), \rho_i(\mathbf{r}))]$, is given by

$$\mathcal{F}[\mathbf{P}_{op}(\mathbf{r}), \mathbf{P}_{in}(\mathbf{r}), \rho_i(\mathbf{r})]$$

$$= \langle \Psi(\mathbf{r})|H_{vac}(\mathbf{r})|\Psi(\mathbf{r})\rangle + \frac{1}{2\chi_0} \int d\mathbf{r}\, \mathbf{P}_{in}(\mathbf{r}) \cdot \mathbf{P}_{in}(\mathbf{r})$$

$$+ \frac{1}{2\chi_{op}} \int d\mathbf{r}\, \mathbf{P}_{op}(\mathbf{r}) \cdot \mathbf{P}_{op}(\mathbf{r})$$

$$+ \frac{1}{2} \iint d\mathbf{r}\, d\mathbf{r}' (\mathbf{P}_{op}(\mathbf{r}) + \mathbf{P}_{in}(\mathbf{r})) \cdot \nabla\nabla' \left(\frac{1}{|\mathbf{r} - \mathbf{r}'|} \right) \cdot (\mathbf{P}_{op}(\mathbf{r}') + \mathbf{P}_{in}(\mathbf{r}'))$$

$$- \int d\mathbf{r}\, (\mathbf{P}_{op}(\mathbf{r}) + \mathbf{P}_{in}(\mathbf{r})) \cdot \mathbf{E}[\mathbf{r}; \rho_i(\mathbf{r})] \tag{14.5}$$

The expectation value of the gas phase Hamiltonian (H_{vac}) is

$$\mathcal{E}^\circ = \langle \Psi(\mathbf{r})|H_{vac}(\mathbf{r})|\Psi(\mathbf{r})\rangle \tag{14.6}$$

and the induced solvent optical and inertial polarization fields interacting with each other and with the electrical field due to the solute charge density, $\rho_i(\mathbf{r})$, are in equilibrium. Therefore it is straightforward to verify that the equilibrium conditions applied to $\mathbf{P}_{in}(\mathbf{r})$, $\mathbf{P}_{op}(\mathbf{r})$, and $\Psi(\mathbf{r})$ lead to the integrals defining the equilibrated polarization vectors [125, 138],

$$\mathbf{P}_{op}^{eq}(\mathbf{r}) = -\chi_{op}\mathbf{E}[\mathbf{r}; \rho_i(\mathbf{r})] - \chi_{op} \int d\mathbf{r}' (\mathbf{P}_{op}^{eq}(\mathbf{r}') + \mathbf{P}_{in}^{eq}(\mathbf{r}')) \cdot \nabla\nabla' \left(\frac{1}{|\mathbf{r} - \mathbf{r}'|} \right) \tag{14.7}$$

$$\mathbf{P}_{in}^{eq}(\mathbf{r}) = -\chi_o\mathbf{E}[\mathbf{r}; \rho_i(\mathbf{r})] - \chi_o \int d\mathbf{r}' (\mathbf{P}_{op}^{eq}(\mathbf{r}') + \mathbf{P}_{in}^{eq}(\mathbf{r}')) \cdot \nabla\nabla' \left(\frac{1}{|\mathbf{r} - \mathbf{r}'|} \right) \tag{14.8}$$

or,

$$\mathbf{P}^{eq}(\mathbf{r}) = \mathbf{P}_{in}^{eq}(\mathbf{r}) + \mathbf{P}_{op}^{eq}(\mathbf{r})$$

$$= -\chi\mathbf{E}[\mathbf{r}; \rho_i(\mathbf{r})] - \chi \int d\mathbf{r}' \, \mathbf{P}^{eq}(\mathbf{r}') \cdot \nabla\nabla' \left(\frac{1}{|\mathbf{r} - \mathbf{r}'|} \right) \tag{14.9}$$

The electric field $\mathbf{E}[\mathbf{r}; \rho_i(\mathbf{r})]$ is produced at $\mathbf{r}$ by the solute charge distribution $\rho_i(\mathbf{r})$, that is,

$$\mathbf{E}[\mathbf{r}; \rho_i(\mathbf{r})] = -\nabla\Phi_i(\mathbf{r}) = -\nabla \int \frac{d\mathbf{r}' \, \rho_i(\mathbf{r}')}{|\mathbf{r} - \mathbf{r}'|} \tag{14.10}$$

The optical (χ_{op}), inertial (χ_{in}) and equilibrium electric susceptibilities (χ_{st}) are related to the static and optical dielectric constants, respectively, as

$$1 + 4\pi\chi_{op} = \epsilon_{op} \tag{14.11}$$

$$\epsilon_{op} + 4\pi\chi_{in} = \epsilon_{st} \tag{14.12}$$

$$1 + 4\pi(\chi_{in} + \chi_{op}) = 1 + 4\pi\chi_{st} = \epsilon_{st} \tag{14.13}$$

Utilizing the Eqs. (14.5), and (14.7–14.9), we determine the stationary value of the free energy to be [125, 138]

$$\mathcal{F}^{eq}[\mathbf{P}^{eq}(\mathbf{r}), \rho_i(\mathbf{r})] = \langle \Psi_i(\mathbf{r})|H_{vac}(\mathbf{r})|\Psi_i(\mathbf{r})\rangle - \frac{1}{2}\int d\mathbf{r}\, \mathbf{P}^{eq}(\mathbf{r}) \cdot \mathbf{E}[\mathbf{r}; \rho_i(\mathbf{r})]$$

$$= \langle \Psi_i(\mathbf{r})|H_{vac}(\mathbf{r})|\Psi_i(\mathbf{r})\rangle + \frac{1}{2}\int d\mathbf{r}\, \rho_i(\mathbf{r})\Phi_i(\mathbf{r}) \qquad (14.14)$$

and furthermore we obtain the nonlinear Schrödinger equation for the solute electronic wave function [125, 138]

$$\left\{ H_{vac}(\mathbf{r}) - \int d\mathbf{r}'(\mathbf{P}_{op}^{eq}(\mathbf{r}) + \mathbf{P}_{in}^{eq}(\mathbf{r})) \cdot \nabla'\left(\frac{1}{|\mathbf{r} - \mathbf{r}'|}\right) \right\} \Psi_i(\mathbf{r}) = \mathcal{E}\Psi_i(\mathbf{r}) \quad (14.15)$$

The nonlinearity of the electronic solute Schrödinger equation occurs due to the implicit dependence of the polarization vector on the solute electronic distribution probability, Eq. (14.9).

A sudden change in the charge distribution of the solute will lead to a shift from a previously reached equilibrium state of the solute-solvent system. One assumes that the optical polarization vector is always in equilibrium with the molecular charge distribution because of its very short relaxation time. In other words the optical polarization field is considered to readjust instantaneously to changes in the solute charge distribution. As mentioned in the introduction to this chapter, the optical polarization represents the response from the quantum degrees of freedom of the outer solvent, such as the electronic degrees of freedom.

The slow-responding polarization, the inertial polarization, describes the response due to the nuclear degrees of freedom of the solvent molecules and the solvent molecular motion—both rotational and translational. Ideally, the inertial polarization should be divided into different types of polarization with significantly different relaxation times, according to the intramolecular vibrations of the solvent molecules, and the rotational and translational motion. Assuming a Debye-type response for the relaxation of the inertial polarization, one finds for most solvents relaxation times that are on the order of nanoseconds to picoseconds [147–149]. With this in mind, it is physically reasonable to assume that during an electronic excitation the inertial polarization remains fixed. Therefore, the inertial polarization vector corresponds to the molecular charge distribution of the initial state. This assumption is completely analogous to that of Franck-Condon and the sudden approximations frequently used in the analysis of electronic spectra for compounds in vacuum.

For the final state, we then have the electronic state of the solute given by the electronic wave function $\Psi_f(\mathbf{r})$, and only the optical polarization is in equilibrium with the charge distribution of the final state. The inertial polarization vector of the final state is that corresponding to the initial state and therefore

$$\mathbf{P}_{op}^{eq}(\mathbf{r}) \equiv \mathbf{P}_{op}^{eq}[\mathbf{r}; \rho_f(\mathbf{r}), \mathbf{P}_{in}(\mathbf{r}; \rho_i(\mathbf{r}))] \tag{14.16}$$

and

$$\mathbf{P}_{in}(\mathbf{r}) = \mathbf{P}_{in}[\mathbf{r}; \rho_i(\mathbf{r})] \tag{14.17}$$

where $\rho_f(\mathbf{r})$ is the charge distribution of the solute in the final state.

The equilibrated optical polarization vector $\mathbf{P}_{op}(\mathbf{r})$ value is determined by the stationary condition with respect to the functional of Eq. [14.5]

$$\frac{\delta}{\delta \mathbf{P}_{op}(\mathbf{r})} \{\mathcal{F}[\mathbf{P}_{op}(\mathbf{r}), \mathbf{P}_{in}(\mathbf{r}), \rho_i(\mathbf{r})]\} = 0 \tag{14.18}$$

and we find [125, 138]

$$\mathbf{P}_{op}^{eq}(\mathbf{r}) = -\chi_{op} \mathbf{E}[\mathbf{r}; \rho_i(\mathbf{r})] - \chi_{op} \int d\mathbf{r}'(\mathbf{P}_{op}^{eq}(\mathbf{r}') + \mathbf{P}_{in}(\mathbf{r}')) \cdot \nabla\nabla'\left(\frac{1}{|\mathbf{r} - \mathbf{r}'|}\right)$$

$$\tag{14.19}$$

The inertial polarization vector does not fulfill an equivalent condition, as it is in a nonequilibrium state with respect to both the optical solvent polarization field and the charge distribution of the solute.

The Schrödinger equation for this nonequilibrated solute–solvent interaction maintains the same form as in Eq. (14.15), with the appropriate redefinition of the equilibrated optical and arbitrary nonequilibrated inertial polarization vectors.

The total contribution to the free energy of the solute-solvent system, for a given initial $\Psi_i(\mathbf{r})$ and final $\Psi_f(\mathbf{r})$ electronic state of the solute reduces to [125, 138]

$$\mathcal{F}^{neq}[\mathbf{P}_{op}^{eq}(\mathbf{r}), \mathbf{P}_{in}(\mathbf{r}), \rho_f(\mathbf{r})] = \langle \Psi_f(\mathbf{r})|H_{vac}(\mathbf{r})|\Psi_f(\mathbf{r})\rangle$$

$$- \frac{1}{2} \int d\mathbf{r} \, \mathbf{P}_{op}^{eq}(\mathbf{r}) \cdot \mathbf{E}[\mathbf{r}; \rho_f(\mathbf{r})]$$

$$- \frac{1}{2} \int d\mathbf{r} \, \mathbf{P}_{in}(\mathbf{r}) \cdot \mathbf{E}[\mathbf{r}; \rho_f(\mathbf{r})]$$

$$+ \frac{1}{2} \int d\mathbf{r} \, \mathbf{P}_{in}(\mathbf{r}) \cdot \left(\frac{\mathbf{P}_{in}(\mathbf{r})}{\chi_{in}} - \frac{\mathbf{P}_{op}^{eq}(\mathbf{r})}{\chi_{op}}\right) \tag{14.20}$$

In order to proceed we let the solute molecule be embedded in a cavity of an effective dimension r_c. For an arbitrary distance $r \gg r_c$ from the origin (cavity center) in the solvent medium we perform a multipole expansion for the solute charge density and the dipole tensor. We then rewrite the nonequilibrium polarization vectors as [138]

$$\mathbf{P}_{\text{op}}^{\text{eq}}(\mathbf{r}) = -\chi_{\text{op}} \sum_{lm} \sqrt{\frac{4\pi}{(2l+1)}} \left[\frac{(2l+1)}{(2l+1) + 4\pi\chi_{\text{op}}(l+1)} \right]$$

$$\left[\langle M_{lm}(\rho_f) \rangle - \frac{4\pi\chi_{\text{in}}(l+1)\langle M_{lm}(\rho_i) \rangle}{(2l+1) + 4\pi\chi_{\text{st}}(l+1)} \right] \nabla[r^{-(l+1)}Y_{lm}^*(\Omega)] \quad (14.21)$$

$$\mathbf{P}_{\text{in}}(\mathbf{r}) = -\chi_{\text{in}} \sum_{lm} \sqrt{\frac{4\pi}{(2l+1)}} \frac{(2l+1)\langle M_{lm}(\rho_i) \rangle}{(2l+1) + 4\pi\chi_{\text{st}}(l+1)} \nabla[r^{-(l+1)}Y_{lm}^*(\Omega)]$$

$$(14.22)$$

where the solute charge multipole density moments $\langle M_{im}(\rho) \rangle$ and the spherical polynomials $S_{l,m}(\mathbf{r})$, are respectively defined to be

$$\langle M_{lm}(\rho) \rangle = \int d\mathbf{r} \, \rho(\mathbf{r}) S_{l,m}(\mathbf{r}) \quad (14.23)$$

$$S_{l,m}(\mathbf{r}) = \sqrt{\frac{4\pi}{(2l+1)}} \, r^l Y_{lm}(\Omega) \quad (14.24)$$

where $Y_{lm}(\Omega)$ is the conventional spherical harmonics and Ω is the polar angles θ and ϕ.

We introduce a partial multipole cavity shape parameter $C_{ll'mm'}$ defined by

$$C_{ll'mm'} = \oint dS(r(\Omega))^{-(l+l'+3)} Y_{lm}^*(\Omega) Y_{l'm'}(\Omega) \quad (14.25)$$

since the cavity enclosing the solute molecule is of a general shape and the partial multipole cavity shape parameters are evaluated knowing the surface's equation $r = r(\Omega)$. For a spherical cavity shape of radius a, the partial multipole cavity shape parameter becomes [138]

$$C_{ll'mm'} = a^{-(2l+1)}\delta_{ll'}\delta_{mm'} \quad (14.26)$$

Utilizing the multipole expansion and allowing for a cavity of general shape, we obtain the following expression for the nonequilibrium free-energy functional

$$\mathcal{F}^{neq}[\Psi, \chi_{in}, \chi_{op}, \rho_f, \rho_i]$$

$$= \langle \Psi_f(\mathbf{r}) | H_{vac}(\mathbf{r}) | \Psi_f(\mathbf{r}) \rangle$$

$$- \frac{4\pi\chi_{op}}{2} \sum_{lm} \sum_{l'm'} \frac{(2l+1) C_{ll'mm'}(l+1) \langle M_{lm}^*(\rho_f) \rangle \langle M_{l'm'}(\rho_f) \rangle}{\sqrt{(2l+1)(2l'+1)}[(2l+1) + 4\pi\chi_{op}(l+1)]}$$

$$- \frac{4\pi\chi_{in}}{2} \sum_{lm} \sum_{l'm'} \frac{(2l+1)^2(l+1) C_{ll'mm'}[2\langle M_{lm}^*(\rho_f) \rangle - \langle M_{lm}^*(\rho_i) \rangle] \langle M_{l'm'}(\rho_i) \rangle}{\sqrt{(2l+1)(2l'+1)}[(2l+1) + 4\pi\chi_{st}(l+1)][(2l+1) + 4\pi\chi_{op}(l+1)]}$$

$$(14.27)$$

In the case of a spherical cavity this reduces to

$$\mathcal{F}^{neq}[\Psi, \chi_{in}, \chi_{op}, \rho_f, \rho_i]$$

$$= \langle \Psi_f(\mathbf{r}) | H_{vac}(\mathbf{r}) | \Psi_f(\mathbf{r}) \rangle$$

$$- \frac{4\pi\chi_{op}}{2} \sum_{lm} \frac{a^{-(2l+1)}(l+1) \langle M_{lm}^*(\rho_f) \rangle \langle M_{lm}(\rho_f) \rangle}{[(2l+1) + 4\pi\chi_{op}(l+1)]}$$

$$- \frac{4\pi\chi_{in}}{2} \sum_{lm} \frac{a^{-(2l+1)}(2l+1)(l+1)[2\langle M_{lm}^*(\rho_f) \rangle - \langle M_{lm}^*(\rho_i) \rangle] \langle M_{lm}(\rho_i) \rangle}{[(2l+1) + 4\pi\chi_{st}(l+1)][(2l+1) + 4\pi\chi_{op}(l+1)]}$$

$$(14.28)$$

where we proceed by substituting the static and optical dielectric constants for the respective electric susceptibilities and rearranging the factors. Thus we obtain [138]

$$\mathcal{F}^{neq}[\Psi, \epsilon_{st}, \epsilon_{op}, \rho_f, \rho_i]$$

$$= \langle \Psi_f(\mathbf{r}) | H_{vac}(\mathbf{r}) | \Psi_f(\mathbf{r}) \rangle + \sum_{lm} g_l(\epsilon_{op}) \langle M_{lm}^*(\rho_f) \rangle \langle M_{lm}(\rho_f) \rangle$$

$$+ \sum_{lm} g_l(\epsilon_{st}, \epsilon_{op})[2\langle M_{lm}^*(\rho_f) \rangle - \langle M_{lm}^*(\rho_i) \rangle] \langle M_{lm}(\rho_i) \rangle \quad (14.29)$$

where the functions $g_l(\epsilon_{st})$ and $g_l(\epsilon_{st}, \epsilon_{op})$ are given by

$$g_l(\epsilon) = -\frac{a^{-(2l+1)}(l+1)(\epsilon-1)}{2[l+\epsilon(l+1)]} \tag{14.30}$$

and

$$g_l(\epsilon_{st}, \epsilon_{op}) = g_l(\epsilon_{st}) - g_l(\epsilon_{op}) \tag{14.31}$$

These functions represent the response of the dielectric medium.

The three terms in Eq. (14.29) are the expectation value of the vacuum Hamiltonian, the solvent energy due to the optical polarization, and the solvent energy contribution arising from the nonequilibrium state of the inertial polarization. The optical polarization vector and the molecular charge distribution $\rho_f(\mathbf{r})$ are in equilibrium with each other. The inertial polarization vector corresponds to the molecular charge distribution $\rho_i(\mathbf{r})$, and it interacts with the molecular charge distribution $\rho_f(\mathbf{r})$, the inertial polarization and $\rho_f(\mathbf{r})$ not being in equilibrium with each other.

Letting the two molecular charge distributions $\rho_f(\mathbf{r})$ and $\rho_i(\mathbf{r})$ be identical, we obtain the usual expression for the equilibrium free energy [117, 133, 134, 150]

$$\mathcal{F}^{eq}[\Psi, \epsilon_{st}, \rho] = \langle\Psi_f(\mathbf{r})|H_{vac}(\mathbf{r})|\Psi_f(\mathbf{r})\rangle + \sum_{lm} g_l(\epsilon_{st})|\langle M_{lm}(\rho)\rangle|^2 \tag{14.32}$$

14.1.2   Implementation

This section describes the implementation of the nonequilibrium solvent model for MCSCRF electronic wavefunctions [138]. We start out by using the total electronic free energy for the nonequilibrium solvent model Eq. (14.29)

$$\mathcal{F}^{neq}(\lambda) = \mathcal{E}_{vac}(\lambda) + \mathcal{E}_{sol}^{neq}(\lambda), \tag{14.33}$$

and obtain an energy functional containing a set of parameters for the electronic wave function λ, which is the combined set of orbital rotational $\{\kappa\}$ and configurational $\{c_i\}$ variational parameters. The expectation value of the vacuum Hamiltonian, $\mathcal{E}_{vac}(\lambda)$, is evaluated from

$$\mathcal{E}_{vac}(\lambda) = \frac{\langle 0|H_{vac}|0\rangle}{\langle 0|0\rangle} \tag{14.34}$$

where a spin-independent Born-Oppenheimer electronic Hamiltonian is introduced

$$H_{\text{vac}} = \sum_{rs} h_{rs} \mathbf{E}_{rs} + \frac{1}{2} \sum_{rstu} (rs|tu) \mathbf{e}_{rs,tu} \tag{14.35}$$

The spin-free operators $\mathbf{E}_{rs}$ and $\mathbf{e}_{rs,tu}$ are written as

$$\mathbf{E}_{rs} = \sum_{\sigma} \mathbf{a}_{r\sigma}^{\dagger} \mathbf{a}_{s\sigma} \tag{14.36}$$

$$\mathbf{e}_{rs,tu} = \sum_{\sigma\sigma'} \mathbf{a}_{r\sigma}^{\dagger} \mathbf{a}_{t\sigma'}^{\dagger} \mathbf{a}_{u\sigma'} \mathbf{a}_{s\sigma} \tag{14.37}$$

where the operators, $\mathbf{a}_{r\sigma}^{\dagger}$ and $\mathbf{a}_{r\sigma}$, are the creation and annihilation operators for an electron in orbital $\phi_r(\mathbf{r})$ with spin σ and with the usual one- and two-electron integrals given by h_{rs} and $(rs|tu)$.

The MCSCF electronic wave function is represented by

$$|0\rangle = \exp\left[\sum_{r>s} \kappa_{sr}(\mathbf{E}_{sr} - \mathbf{E}_{rs})\right] \sum_i c_i |\Theta_i\rangle \tag{14.38}$$

where $\{\kappa_{sr}\}$ and $\{c_i\}$ are the orbital rotational and configurational variation parameters, respectively, and $|\Theta_i\rangle$ denotes the set of configuration state functions (CSFs). The wave function is determined by locating the stationary value of λ corresponding to Eq. (14.33). This stationary point is characterized by a vanishing gradient and a correct Hessian index of $\mathcal{F}^{\text{neq}}(\lambda)$.

The next step is to expand the energy functional to second order in the non-redundant variables around a given set of $\lambda^{(k)}$ values, where k is the iteration number

$$\Delta E^{(2)}(\lambda - \lambda^{(k)}; \lambda^{(k)}) = \mathcal{E}^{(2)}(\lambda - \lambda^{(k)}; \lambda^{(k)}) - \mathcal{E}(\lambda^{(k)})$$

$$= \mathbf{g}^T(\lambda - \lambda^{(k)}) + \tfrac{1}{2}(\lambda - \lambda^{(k)})^T \mathbf{H}(\lambda - \lambda^{(k)}) \tag{14.39}$$

and where $\mathbf{g}$ and $\mathbf{H}$ are the gradient and Hessian of $\mathcal{F}^{\text{neq}}(\lambda)$ at $\lambda^{(k)}$

The nonequilibrium solvation energy, Eq. (14.29), is given by

$$\mathcal{E}_{\text{sol}}^{\text{neq}} = \sum_{lm} R_{lm}(\rho_f, \epsilon_{\text{op}}) \langle T_{lm}(\rho_f) \rangle$$

$$+ \sum_{lm} R_{lm}(\rho_i, \epsilon_{\text{st}}, \epsilon_{\text{op}}) [2\langle T_{lm}(\rho_f) \rangle - \langle T_{lm}(\rho_i) \rangle] \tag{14.40}$$

where

$$R_{lm}(\rho_f, \epsilon_{op}) = g_l(\epsilon_{op}) \langle T_{lm}(\rho_f) \rangle$$

$$R_{lm}(\rho_i, \epsilon_{st}, \epsilon_{op}) = g_l(\epsilon_{st}, \epsilon_{op}) \langle T_{lm}(\rho_i) \rangle \qquad (14.41)$$

and where the $\{\langle T_{lm} \rangle\}$ elements are the expectation values of the nuclear and electronic solvent operators

$$\langle T_{lm} \rangle = \langle T_{lm}^n \rangle - \langle T_{lm}^e \rangle$$

$$T_{lm}^n = \sum_g Z_g t_{lm}(\mathbf{R}_g)$$

$$T_{lm}^e = t_{lm}(\mathbf{r}) = \sum_{pq} t_{lm}^{pq} \mathbf{E}_{pq} \qquad (14.42)$$

Here Z_g and $\mathbf{R}_g$ are the charge and position vector of nucleus g, respectively. The subscripts p and q represent the orbitals ϕ_p and ϕ_q. We have defined $t_{lm}(\mathbf{R}_q)$ and t_{lm}^{pg} through the functions

$$t_{l,0}(\mathbf{r}) = S_{l,0}(\mathbf{r})$$

$$t_{l,m}(\mathbf{r}) = \frac{\sqrt{2}}{2}(S_{l,m}(\mathbf{r}) + S_{l,-m}(\mathbf{r}))$$

$$t_{l,-m}(\mathbf{r}) = \frac{\sqrt{2}}{2i}(S_{l,m}(\mathbf{r}) - S_{l,-m}(\mathbf{r})) \qquad (14.43)$$

where m is a positive integer, and the functions, $S_{l,m}(\mathbf{r})$, are the conventional spherical polynomials.

14.1.3   Optimization Procedure

In this section we consider the evaluation of the gradient and Hessian terms that are needed for utilizing Eq. (14.39) [138].

The contribution to the gradient due to the solvation energy in Eq. (14.29) is given as

$$\frac{\partial E_{sol}}{\partial \lambda_i} = \sum_{lm} 2g_l(\epsilon_{op})\langle T_{lm}(\rho_f)\rangle \frac{\partial \langle T_{lm}(\rho_f)\rangle}{\partial \lambda_i}$$

$$+ \sum_{lm} 2g_l(\epsilon_{st}, \epsilon_{op})\langle T_{lm}(\rho_i)\rangle \frac{\partial \langle T_{lm}(\rho_f)\rangle}{\partial \lambda_i}$$

$$= -\sum_{lm} 2g_l(\epsilon_{op})\langle T_{lm}(\rho_f)\rangle \frac{\partial \langle T^e_{lm}(\rho_f)\rangle}{\partial \lambda_i}$$

$$- \sum_{lm} 2g_l(\epsilon_{st}, \epsilon_{op})\langle T_{lm}(\rho_i)\rangle \frac{\partial \langle T^e_{lm}(\rho_f)\rangle}{\partial \lambda_i} \qquad (14.44)$$

where the electronic contribution is obtained as

$$\langle T^e_{lm}(\rho_f)\rangle = \frac{\langle 0|T^e_{lm}(\rho_f)|0\rangle}{\langle 0|0\rangle} = \sum_{pq} D_{pq} t^{pq}_{lm} \qquad (14.45)$$

and

$$t^{pq}_{lm} = \langle \phi_p(\mathbf{r})|t_{lm}(\mathbf{r})|\phi_q(\mathbf{r})\rangle \qquad (14.46)$$

and where D_{pq} is the one-electron density matrix. We introduce the effective one-electron operator

$$T^g = -2\sum_{lm} [g_l(\epsilon_{op})\langle T_{lm}(\rho_f)\rangle + g_l(\epsilon_{st}, \epsilon_{op})\langle T_{lm}(\rho_i)\rangle] t_{l,m}(\mathbf{r}) \qquad (14.47)$$

and obtain the configuration and orbital parts of the gradient [138]

$$\frac{\partial E_{sol}}{\partial c_\mu} = -\sum_{lm} 2[g_l(\epsilon_{op})\langle T_{lm}(\rho_f)\rangle + g_l(\epsilon_{st}, \epsilon_{op})\langle T_{lm}(\rho_i)\rangle] \frac{\partial \langle T^e_{lm}(\rho_f)\rangle}{\partial c_\mu}$$

$$= 2[\langle \mu|T^g|0\rangle - \langle 0|T^g|0\rangle c_\mu] \qquad (14.48)$$

and

$$\frac{\partial E_{\text{sol}}}{\partial \kappa_{rs}} = -\sum_{lm} 2[g_l(\epsilon_{\text{op}})\langle T_{lm}(\rho_f)\rangle + g_l(\epsilon_{\text{st}}, \epsilon_{\text{op}})\langle T_{lm}(\rho_i)\rangle] \frac{\partial \langle T^e_{lm}(\rho_f)\rangle}{\partial \kappa_{rs}}$$

$$= 2\langle 0|[\mathbf{E}_{rs}, T^g]|0\rangle \tag{14.49}$$

The effective one-electron operator T^g is constructed utilizing the electronic wave function currently being optimized and the electronic wave function of a different state, such as the initial state in an electronic excitation.

We will now consider the contribution to the Hessian due to the solvation energy in Eq. (14.29). The direct, restricted-step, second order MCSCF strategy [151], involves a linear transformation algorithm constructing directly from trial vectors $\mathbf{b}^{(k)}$ the Hessian contribution to the energy functional

$$\sigma_i^{(k)} = \sum_j \frac{\partial^2 E}{\partial \lambda_j \partial \lambda_i} \mathbf{b}_i^{(k)} \tag{14.50}$$

For the case of the CSF trial vector we have the following contribution due to the nonequilibrium solvent energy [138]

$$\sigma_j^{c,\text{sol}} = \sum_\nu \frac{\partial^2 E_{\text{sol}}}{\partial c_\nu \partial \lambda_j} \mathbf{b}^{(c)}$$

$$= -\sum_{lm} 2[g_l(\epsilon_{\text{op}})\langle T_{lm}(\rho_f)\rangle + g_l(\epsilon_{\text{st}}, \epsilon_{\text{op}})\langle T_{lm}(\rho_i)\rangle] \sum_\nu \frac{\partial^2 \langle T^e_{lm}(\rho_f)\rangle}{\partial c_\nu \partial \lambda_j} \mathbf{b}_\nu^{(c)}$$

$$+ \sum_{lm} 2g_l(\epsilon_{\text{op}}) \frac{\partial \langle T^e_{lm}\rangle}{\partial \lambda_j} \sum_\nu \frac{\partial \langle T^e_{lm}\rangle}{\partial c_\nu} \mathbf{b}_\nu^{(c)} \tag{14.51}$$

which gives

$$\sigma_\mu^{c,\text{sol}} = 2[\langle \mu|T^g|\mathbf{B}\rangle - \langle 0|T^g|0\rangle \mathbf{b}_\mu^{(c)}] + 2[\langle \mu|T^{xc}|0\rangle - \langle 0|T^{xc}|0\rangle c_\mu] \tag{14.52}$$

and

$$\sigma_{rs}^{c,\text{sol}} = 2[\langle 0|[\mathbf{E}_{rs}, T^g]|\mathbf{B}\rangle + \langle 0|[\mathbf{E}_{rs}, T^{xc}]|\mathbf{B}\rangle] \tag{14.53}$$

Here we have introduced

$$\mathbf{T}^{xc} = 4 \sum_{lm} g_l(\epsilon_{op}) \langle 0|t_{lm}|B\rangle \, t_{lm} \tag{14.54}$$

and

$$|B\rangle = \sum_{\mu} \mathbf{b}_{\mu}^{(c)} |\Theta_{\mu}\rangle \tag{14.55}$$

The contribution to the linear transformation of an orbital trial vector is given by [138]

$$
\begin{aligned}
\sigma_j^{o,\,sol} &= \sum_{sr} \frac{\partial^2 E_{sol}}{\partial \lambda_j \partial \kappa_{sr}} \, \mathbf{b}_{sr}^{(o)} \\
&= -\sum_{lm} 2[g_l(\epsilon_{op})\langle T_{lm}(\rho_f)\rangle + g_l(\epsilon_{st},\epsilon_{op})\langle T_{lm}(\rho_i)\rangle] \sum_{sr} \frac{\partial^2 \langle T_{lm}^e(\rho_f)\rangle}{\partial \lambda_j \partial \kappa_{sr}} \, \mathbf{b}_{sr}^{(o)} \\
&\quad + \sum_{lm} 2 g_l(\epsilon_{op}) \frac{\partial \langle T_{lm}^e\rangle}{\partial \lambda_j} \sum_{rs} \frac{\partial \langle T_{lm}^e\rangle}{\partial \kappa_{rs}} \, \mathbf{b}_{sr}^{(o)} \tag{14.56}
\end{aligned}
$$

We find that

$$\sigma_j^{o,\,sol} = 2\langle \mu|T^{yo}|0\rangle + 2[\langle \mu|T^{xo}|0\rangle - \langle 0|T^{xo}|0\rangle c_{\mu}] \tag{14.57}$$

and

$$
\begin{aligned}
\sigma_{rs}^{o,sol} &= 2\langle 0|[\mathbf{E}_{rs}, T^{yo}]|0\rangle + 2\langle 0|[\mathbf{E}_{rs}, T^{xo}]|0\rangle \\
&\quad + \sum_{t} (\langle 0|[\mathbf{E}_{ts}, T^{g}]|0\rangle \mathbf{b}_{rt} - \langle 0|[\mathbf{E}_{tr}, T^{g}]|0\rangle \mathbf{b}_{st}) \tag{14.58}
\end{aligned}
$$

The effective operators T^{yo} and T^{xo} are defined as

$$T^{yo} = -2 \sum_{lm} [g_l(\epsilon_{op})\langle T_{lm}(\rho_f)\rangle + g_l(\epsilon_{st},\epsilon_{op})\langle T_{lm}(\rho_i)\rangle] Q^{lm} \tag{14.59}$$

and

$$T^{xo} = 2 \sum_{lm} g_l(\epsilon_{op})\langle 0|Q^{lm}|0\rangle t^{lm} \qquad (14.60)$$

where we have introduced the one-index tranformed solvent integrals, Q^{lm}, generated by the orbital trial vector [134, 138, 152, 153]. With the analytical formulation of the first and second order derivatives of the electronic energy with respect to the solvent interaction we thus find that the optimization of the reaction field and the vacuum wave functions show similar characteristics. The implication of this is that the computational effort of the reaction field calculations is only marginally larger than of the ordinary Hartree-Fock or MCSCF vacuum calculations; only an extra storage and retrieval of the one-electron solvent integrals is required.

14.2 NUMERICAL EXAMPLES

As a numerical example of the method we consider the excitation energies of H_2CO solvated in various solvents and in vacuum [138]. We perform MCSCRF calculations on the ground state (GS) and on the three different excited states (the B_1, A_2, and B_2 excited states of H_2CO). The calculations for the ground state of H_2CO in the different solvents were performed utilizing the equilibrium expression for the energy functional in Eq. (14.32) since the solvent, in the initial state of the excitation process, is in equilibrium with the actual charge distribution of the solute, the one corresponding to the ground state of H_2CO. The final state of the excitation process leads to a nonequilibrium solvent configuration, since the inertial polarization is unable to change during the short time period of the vertical excitation process. The calculations of the final states are performed utilizing the nonequilibrium expression for the energy functional in Eq. (14.29). In Table 14.1 we present results of the excitation energies of

TABLE 14.1. MCSCRF Calculations of Excitation Energies of H_2CO for the B_1, A_2, and B_2 Excited States

Vacuum or Solvent	GS → B_1 (eV)	GS → A_2 (eV)	GS → B_2 (eV)
Vacuum	9.909	7.613	4.473
Benzene	9.949	7.351	4.491
Ethylether	10.001	7.736	4.517
1-Hexanol	10.038	7.847	4.534
Acetone	10.050	7.920	4.540
Methanol	10.057	7.956	4.544
Water	10.063	7.971	4.546

TABLE 14.2. The Contribution to the Dielectric Solvation Energy for the Different Multipoles[a]

l	GS	A_2	B_1	B_2
0	0.000000	0.000000	0.000000	0.000000
1	−0.005365	0.008491	0.001022	−0.001812
2	0.000000	−0.000343	−0.000161	−0.000242
3	−0.000099	0.000153	−0.000252	−0.000264
4	−0.000024	0.000004	−0.000080	−0.000189
5	−0.000032	0.000015	−0.000060	−0.000086
6	−0.000011	0.000001	−0.000018	−0.000028
7	−0.000007	0.000002	−0.000009	−0.000012

[a]The ground state (GS) and the excited states (A_2, B_1, and B_2) are analyzed. The multipole expansion is given by the multipole parameter l.

H_2CO for the excitations to the B_1, A_2, and B_2 excited states. The calculations were performed

(1) Within the C_{2v} point group
(2) For the solvents or dielectric media:
   Benzene ($\epsilon_{st} = 2.284$, $\epsilon_{op} = 2.244$)
   Ethyl ether ($\epsilon_{st} = 4.335$, $\epsilon_{op} = 1.828$)
   1-Hexanol ($\epsilon_{st} = 13.3$, $\epsilon_{op} = 2.005$)
   Acetone ($\epsilon_{st} = 20.7$, $\epsilon_{op} = 1.841$)
   Methanol ($\epsilon_{st} = 32.63$, $\epsilon_{op} = 1.758$)
   Water ($\epsilon_{st} = 78.54$, $\epsilon_{op} = 1.778$)
(3) With a spherical cavity having a radius of 2.645 Å
(4) With a complete active space with 12 active electrons for each of the four states:
   GS: $6a_1, 4b_2, 2b_1, 0a_2$
   A_2: $7a_1, 3b_2, 2b_1, 0a_2$
   B_1: $5a_1, 4b_2, 3b_1, 0a_2$
   B_2: $6a_1, 3b_2, 3b_1, 0a_2$
(5) With a molecular geometry from Herzberg [154] and a basis set from ref. [155]

The results in Table 14.1 give a blue shift for all the excitation processes compared to the excitation energy in vacuum. In Table 14.2 we show how the different terms in the multipole expansion contribute to the ground state and the excited states, and it is clearly seen that for the ground state, by far the most important term is $l = 1$, whereas for the excited states the contributions from other terms become larger.

15

MOLECULAR PROPERTIES OF SOLVATED MOLECULES

This chapter addresses the problem of solving time-dependent quantum mechanical equations for a molecular system coupled to a solvent [111, 113, 156–159]. We consider a response formulation for solving the time-dependent response equations when the solvated molecule is perturbed by an external perturbation, such as an electromagnetic field. We begin by defining the necessary building blocks such as the Hamiltonian, the equation of motion, and the response functions. Later we present some of the problems involved when comparing experimental and theoretical investigations. Finally, we outline the mathematical steps involved in determining the molecular properties of a solvated system.

The advantages of this method compared to more conventional approaches are:

(1) A more appropriate representation of the charge distribution of the solute

(2) A scheme for obtaining excitation energies, transition moments, polarizabilities, hyperpolarizabilities, etc., in one set of reaction field response calculations

(3) An incorporation of the total Hamiltonian into the response equations

(4) Handling of both nonequilibrium and equilibrium solvent configurations

The Onsager reaction field model assumes that the dipole moment represents appropriately the solute's charge distribution, which is sufficient for solutes without charge monopoles and higher charge moments than the dipole moment. The strength of response theory lies in the possibility of obtaining many molecular properties in one single response calculation, such as excitation energies,

170

transition moments, frequency-dependent and -independent polarizabilities and hyperpolarizabilities. The incorporation of the solvent interactions into the total Hamiltonian gives rise to new methods and calculations of time-dependent molecular properties of solvated compounds.

15.1 DEFINITIONS

The Hamiltonian for the response of the solvated system (being in a state $|0\rangle$) to an external field is given by

$$H = H_0 + W + V(t) \tag{15.1}$$

where H_0 is the Hamiltonian of an isolated molecular system, and W is the interaction operator describing the interaction between the molecular system and the surrounding medium. The interaction between the solvated system and the external field can, in the frequency domain, be expressed as

$$V(t) = \int_{-\infty}^{+\infty} d\omega \, V^\omega \, \exp(-i\omega t) \tag{15.2}$$

In the gas phase the Hamiltonian is written as

$$H = H_0 + V(t) \tag{15.3}$$

and we make the assumption that the eigenvalues and eigenfunctions of H_0 are known for the reference state

$$H_0|0\rangle = E_0|0\rangle \tag{15.4}$$

and for the complementary set of eigenstates

$$H_0|n\rangle = E_n|n\rangle \tag{15.5}$$

The set of eigenstates $\{|n\rangle\}$ is orthogonal to $|0\rangle$. We determine the time evolution of $|0\rangle$ and write the time-dependent state $|\bar{0}\rangle$ as in ref. [160]

$$|\tilde{0}\rangle = \exp(i\,U(t))|0\rangle \tag{15.6}$$

where $U(t)$ is given by

$$U(t) = \sum_n [U_n |n\rangle \langle 0| + U_n^* |0\rangle \langle n|] \tag{15.7}$$

Ehrenfest's theorem is required to hold for the time-dependent state $|\tilde{0}\rangle$ for the excitation-deexcitation operators Ω, which are elements of the set $\{|n\rangle \langle 0|; |0\rangle \langle n|\}$. Thus,

$$\langle \tilde{0}|\Omega|\dot{\tilde{0}}\rangle + \langle \dot{\tilde{0}}|\Omega|\tilde{0}\rangle = -i \langle \tilde{0}|[\Omega, H_0 + V(t)]|\tilde{0}\rangle \tag{15.8}$$

The general idea, as shown in ref. [160], is to expand $U(t)$ in a power series with respect to $V(t)$

$$U(t) = U(t)^{(1)} + U(t)^{(2)} + \cdots \tag{15.9}$$

The expansion of the exponential generates a power series expansion of the wave function and we will require Ehrenfest's theorem to hold for the time-dependent wave function in all powers of the perturbation.

The definition of the response functions is given by the expansion coefficients of the expectation value of an operator B

$$B_{av}(t) = \langle \tilde{0}|B|\tilde{0}\rangle = \langle 0|B|0\rangle + \int_{-\infty}^{+\infty} d\omega_m \ll B; V^{\omega m} \gg_{\omega_m} \exp(-i\omega_m t)$$

$$+ \frac{1}{2} \int_{-\infty}^{+\infty} d\omega_m \int_{-\infty}^{+\infty} d\omega_n \ll B; V^{\omega m}, V^{\omega n} \gg_{\omega_n + i\epsilon, \omega_n} \cdot \exp[(-i\omega_m - i\omega_n)t] + \cdots$$

$$\tag{15.10}$$

Introduction of the intermediate state $|p\rangle$ allows us to express the linear response function as

$$\langle\langle B; V^{\omega m}\rangle\rangle_{\omega m} = \sum_{p>0} \left\{ \frac{\langle 0|B|p\rangle \langle p|V^{\omega m}|0\rangle}{\omega_m - \omega_{p0}} - \frac{\langle 0|V^{\omega m}|p\rangle \langle p|B|0\rangle}{\omega_m + \omega_{p0}} \right\} \tag{15.11}$$

where the energy difference between the excited state $|p\rangle$ and the ground state $|0\rangle$ is given by $\omega_{p0} = E_p - E_0$, and where ω_m is the frequency of electric field. Similarly we write the quadratic response function as

$$\langle\langle B; V^{\omega_m}, V^{\omega_n}\rangle\rangle_{\omega_m, \omega_n} = \sum_{p,q,>0} \left\{ \frac{\langle 0|B|p\rangle[\langle p|V^{\omega_m}|n\rangle - \delta_{pq}\langle 0|V^{\omega_m}|0\rangle]\langle q|V^{\omega_n}|0\rangle}{(\omega_m + \omega_n - \omega_{p0})(\omega_n - \omega_{q0})} \right.$$

$$+ \frac{\langle 0|V^{\omega_n}|q\rangle[\langle q|V^{\omega_n}p\rangle - \delta_{pq}\langle 0|V^{\omega_m}|0\rangle]\langle p|B|0\rangle}{(\omega_m + \omega_n + \omega_{p0})(\omega_n + \omega_{q0})}$$

$$- \frac{\langle 0|V^{\omega_m}|p\rangle[\langle p|B|q\rangle - \delta_{pq}\langle 0|B|0\rangle]\langle q|V^{\omega_n}|0\rangle}{(\omega_m + \omega_{p0})(\omega_n - \omega_{q0})}$$

$$+ \frac{\langle 0|B|p\rangle[\langle p|V^{\omega_n}|q\rangle - \delta_{pq}\langle 0|V^{\omega_n}|0\rangle]\langle q|V^{\omega_m}|0\rangle}{(\omega_m + \omega_n - \omega_{p0})(\omega_m - \omega_{q0})}$$

$$+ \frac{\langle 0|V^{\omega_m}|q\rangle[\langle q|V^{\omega_n}|p\rangle - \delta_{pq}\langle 0|V^{\omega_n}|0\rangle]\langle p|B|0\rangle}{(\omega_m + \omega_n + \omega_{p0})(\omega_m + \omega_{q0})}$$

$$\left. - \frac{\langle 0|V^{\omega_n}|p\rangle[\langle p|B|q\rangle - \delta_{pq}\langle 0|B|0\rangle]\langle q|V^{\omega_m}|0\rangle}{(\omega_n + \omega_{p0})(\omega_m - \omega_{q0})} \right\}$$

$$(15.12)$$

where ω_m and ω_n are the frequencies of electric field.

We will now consider the interaction operator and illustrate its dependence on the electromagnetic field. For a homogeneous periodic field we are able to write the interaction operator as

$$V^t = -(\mu F^0) \cos \omega t \qquad (15.13)$$

where $\mu = (\mu^x, \mu^y, \mu^z)$ represents the electric dipole moment operator of the molecular system. The electric field is given by

$$F(t) = \tfrac{1}{2} F^0[\exp(-i\omega t) + \exp(i\omega t)] = F^0 \cos \omega t \qquad (15.14)$$

where ω and F^0 are the frequency and electric field strength of the electric field, respectively. The operator V^ω is then given as

$$V^\omega = -\tfrac{1}{2}(\mu F^0)[\delta(\omega - \omega_0) + \delta(\omega + \omega_0)] \qquad (15.15)$$

The quantity μ_i^{ind} is the total dipole moment induced in the molecule by the electric field $F(t)$. It is expressed as a Taylor expansion with respect to the electric field

$$\mu_i^{\text{ind}} = \mu_i^T + \sum_j \alpha_{ij}^T(-\omega;\omega)F_j^\omega + \frac{1}{2}\sum_{j,k} \beta_{ijk}^T(-2\omega;\omega,\omega)F_j^\omega F_k^\omega$$

$$+ \frac{1}{6}\sum_{j,k,l} \gamma_{ijkl}^T(-3\omega;\omega,\omega,\omega)F_j^\omega F_k^\omega F_l^\omega + \cdots \qquad (15.16)$$

The molecular properties of the molecular system are defined through the expansion coefficients μ_i^T, α_{ij}^T, β_{ijk}^T, and γ_{ijkl}^T, which correspond to the dipole moment, frequency-dependent polarizability, first hyperpolarizability, and second hyperpolarizability, respectively. The subscripts i, j, k, and l denote the molecular axes x, y, and z, and the component of the interacting field $\mathbf{F}(t)$ at the frequency ω is given by F_i^ω.

The polarizability and hyperpolarizabilities can be expressed in terms of response functions as

$$\alpha_{ij}^T(-\omega_\delta;\omega_m) = -\langle\langle\mu_i;\mu_j\rangle\rangle_{\omega m} \qquad\qquad \omega_\delta = \omega_m$$

$$\beta_{ijk}^T(-\omega_\delta;\omega_m,\omega_n) = -\langle\langle\mu_i;\mu_j,\mu_k\rangle\rangle_{\omega m,\omega n} \qquad\qquad \omega_\delta = \omega_m + \omega_n$$

$$\gamma_{ijkl}^T(-\omega_\delta;\omega_m,\omega_n,\omega_p) = -\langle\langle\mu_i;\mu_j,\mu_k,\mu_l\rangle\rangle_{\omega m,\omega n,\omega p} \qquad\qquad \omega_\delta = \omega_m + \omega_n + \omega_p$$

$$(15.17)$$

The expressions for the polarizability and hyperpolarizabilities of the solvated system contain a summation over a complete set of exact states. However, the explicit reference to and the summation over the excited states is not the most convenient method with respect to implementation. The actual strategy is to avoid any explicit reference to excited states and perform, instead of a summation over states, iterative solution of linear equations.

Molecular properties of solutes have mainly been measured in two-component systems consisting of solute molecules and solvent molecules. A strategy for performing these measurements was put forward by Singer and Garito [161]. They proposed an infinite dilution extrapolation procedure for determining the solute molecular polarizability and hyperpolarizabilities. Therefore, a single solute molecule is assumed to be completely surrounded by the solvent molecules and we can safely assume that the molecular properties can be evaluated by the system Hamiltonian, $H_0 + W$. Application of the external field to the solution leads to the induction of a field in the solvent molecules and the resulting field is therefore thought of as a screened field (the combined internal and external field) that the solute molecules experience. In general one can characterize such solvent effects as:

(1) Screening of the externally applied field

(2) Being part of the unperturbed molecular Hamiltonian and therefore included in the response equations

The first point, the screening of the external field, is usually taken into account by incorporating into the experimentally observed molecular properties local field factors, see Section 15.2. Point 2 is addressed in the section on solvent-induced molecular properties, see Section 15.4. Local field effects do not take account of the effects of the intermolecular interactions that arise when a molecule is solvated, but do however describe how the external field is screened by the solvent molecules. The interaction between a molecule and a solvent determines the difference between molecular properties in the gas and liquid phases and thus must be included in a quantum description of molecular properties.

15.2   SCREENING OF THE EXTERNALLY APPLIED FIELD. LOCAL FIELD CORRECTIONS

Experimentally one measures the response of the system under investigation to the external electric field. The total polarization P_i ($i = x, y, z$) can be expressed as a power series in the external field strength $\mathbf{F}(\omega)$. Application of a Taylor series leads to the following expression

$$\overline{P}_i = \sum_j \chi_{ij}^{(1)} \mathbf{F}_j^\omega + \frac{1}{2} \sum_{jk} \chi_{ijk}^{(2)} \mathbf{F}_j^\omega \mathbf{F}_k^\omega + \frac{1}{6} \sum_{jkl} \chi_{ijkl}^{(3)} \mathbf{F}_j^\omega \mathbf{F}_k^\omega \mathbf{F}_l^\omega + \cdots \qquad (15.18)$$

where $\chi^{(i)}$ is the susceptibility tensor of rank $(i + 1)$. The susceptibility tensor is connected with the molecular microscopic polarizabilities and in the case of a second harmonic generation (SHG) measurement, one has

$$\overline{P}_i(2\omega) = \frac{1}{2} \sum_{jk} \chi_{ijk}^{(2)}(-2\omega; \omega, \omega) \mathbf{F}_j^\omega \mathbf{F}_k^\omega \qquad (15.19)$$

It is clearly seen that it is necessary to measure the total polarization P_i and the strength of the external field when determining the value of the electric molecular properties.

For the investigated system, solvent and solute, one assumes that it consists of polarizable entities and the result of the applied external electric field leads to a total polarization per unit volume. The total polarization per unit volume is expressed as a sum of effective dipoles induced by the electric field acting at the position of the entities. These induced dipoles interact with each other and this leads to a screening of the externally applied field. The electric field acting on each effective dipole is not rigorously the external field but instead the vector sum of the external field and the fields from the induced dipoles. This effective field is termed the local field. The relations between the macroscopic polarizabilities (susceptibilities) and the microscopic polarizabilities are assumed to be given by [162, 163]

$$\chi^{(1)}(-\omega; \omega) = N s^{(1)}(-\omega; \omega) \overline{\alpha}(-\omega; \omega) \tag{15.20}$$

$$\chi^{(2)}(-\omega_m - \omega_n; \omega_m, \omega_n) = N s^{(2)}(-\omega_m - \omega_n; \omega_m, \omega_n) \tag{15.21}$$

$$\times \overline{\beta}(-\omega_m - \omega_n; \omega_m, \omega_n) \tag{15.22}$$

$$\chi^{(3)}(-\omega_m - \omega_n - \omega_l; \omega_m, \omega_n, \omega_l) = N s^{(3)}(-\omega_m - \omega_n - \omega_l; \omega_m, \omega_n, \omega_l) \tag{15.23}$$

$$\times \overline{\gamma}(-\omega_m - \omega_n - \omega_l; \omega_m, \omega_n, \omega_l) \tag{15.24}$$

where N is the number density, and $s^{(n)}$ represents a local field factor that accounts for the screening of the external field. The orientationally averaged microscopic molecular properties determined in experiments are $\overline{\alpha}$, $\overline{\beta}$, and $\overline{\gamma}$.

In order to make comparisons between experimental measurements and theoretical calculations one must consider local field factors. Investigations of gas phase properties involve local field factors that are approximately equal to one. Experimentally determined microscopic molecular properties can therefore be directly compared to results from theoretical calculations in the gas phase. In the case of investigations concerning molecular properties in the liquid phase, it is important to note that experimentally determined microscopic molecular properties should be compared to the results of the theoretical calculations in liquid phase.

Determination of local field factors involves, in most cases, the application of the Lorentz model [164]. Upon such application one obtains

$$s^{(1)}(-\omega; \omega) = s^\omega s^\omega \tag{15.25}$$

$$s^{(2)}(-\omega_m - \omega_n; \omega_m, \omega_n) = s^{\omega_m + \omega_n} s^{\omega_m} s^{\omega_n} \tag{15.26}$$

$$s^{(3)}(-\omega_m - \omega_n - \omega_l; \omega_m, \omega_n, \omega_l) = s^{\omega_m + \omega_n + \omega_l} s^{\omega_m} s^{\omega_n} s^{\omega_l} \tag{15.27}$$

where

$$s^\omega = \tfrac{1}{3}(n_\omega^2 + 2) \tag{15.28}$$

and where n_ω is the index of refraction of the solvent at the frequency ω of the externally applied field. Equations (15.25–15.28) are called the Lorentz local field factors. In the static limit (the case of an external frequency-independent electromagnetic field), the Onsager expression is often used to describe the local field factor

$$s^0 = \frac{\epsilon(n^2 + 2)}{(n^2 + 2\epsilon)} \tag{15.29}$$

where ϵ is the dielectric constant and n is the index of refraction.

As an example of the tabulation of the experimental results let us consider hyperpolarizabilities measured by the electrical-field-induced second harmonic generation (ESHG), where the field-induced dipole moment is determined at the second harmonic frequency 2ω produced by fields at frequency ω and zero. The field-induced dipole moment becomes (using a Taylor series expansion)

$$\mu_i^{\text{ind}}(2\omega) = \tfrac{1}{4}\beta_{ijk}^T(-2\omega;\omega,\omega)F_j^\omega F_k^\omega + \tfrac{1}{4}\gamma_{ijkl}^T(-2\omega;\omega,\omega,0)F_j^\omega F_k^\omega F_l^0 + \cdots \tag{15.30}$$

The measured polarization $P_i^{2\omega}$ contains contributions from the first and second hyperpolarizabilities and is given by

$$
\begin{aligned}
P_i^{2\omega} &= \frac{1}{4}\, N s^{(3)}(-2\omega;\omega,\omega,0)(\bar{\gamma}^T(-2\omega;\omega,\omega,0) + \frac{\mu_0 \beta_{\parallel}^T(-2\omega;\omega,\omega)}{3kT}\,(F_i^\omega)^2 F_i^0 \\
&= \frac{1}{4}\, N s^{(3)}(-2\omega;\omega,\omega,0)(\bar{\gamma}^T(-2\omega;\omega,\omega,0) + \frac{\mu_0 \beta_z^T(-2\omega;\omega,\omega)}{5kT}\,(F_i^\omega)^2 F_i^0
\end{aligned}
\tag{15.31}
$$

where thermal averaging with respect to the orientations of a dipolar molecule in a static electric field [165] has been performed. The component $\beta_{\parallel}^T$ is the vector component of the third order tensor β_{ijk}^T directed along the dipole moment,

$$\beta_{\parallel}^T = \frac{1}{5}\sum_j (\beta_{zjj}^T + \beta_{jjz}^T + \beta_{jzj}^T) \qquad j = x, y, z \tag{15.32}$$

and β_z^T denotes the z-component of the vector and the following relation holds, $\beta_{\parallel}^T = \frac{3}{5}\beta_z^T$. The term $\bar{\gamma}^T$ is the scalar component of the fourth order tensor γ_{ijkl}

$$\bar{\gamma}^T = \frac{1}{15}\sum_{jk} (\gamma_{jjkk}^T + \gamma_{jkkj}^T + \gamma_{jkjk}^T) \qquad j, k = x, y, z \tag{15.33}$$

The local field factor is as defined in Eq. (15.27)

$$s^{(3)}(-2\omega;\omega,\omega,0) = s^{2\omega}(s^{\omega})^2 s^0 \tag{15.34}$$

15.3 RESPONSE FUNCTIONS IN THE MCSCF APPROXIMATION

The response of a molecular system interacting with a time-dependent field is conveniently described by response functions. We let an observable be represented by a Hermitian operator and solve the equation of motion to a given order in the perturbation. Solving the response of the molecular system to first order in the perturbation leads to the linear response function, the quadratic response gives the quadratic response function, etc. Letting the observable be the induced dipole moment, the electric dipole operator is the Hermitian operator. The results of the response calculation would, in the case where a periodic electric field is the perturbation, be the frequency-dependent polarizability (from the linear response function), the frequency-dependent first hyperpolarizability (from the quadratic response function), and so forth for the higher order responses. In addition one obtains the residues of the response functions that determine transition matrix elements. In our example, the residue of the linear response function determines the one-photon transition moments, and the residues of the quadratic response function determine the two-photon transition moments.

The response function formalism that forms the basis for the approach outlined in this chapter originates from Zubarev [166]. Initially we will outline the development for response methods for molecular properties in a vacuum and subsequently move on to the liquid phase. Response methods involving specific electronic structure models in a vacuum started out with SCF linear and quadratic response functions [167]. Later Dalgaard presented an elegant derivation of the SCF quadratic response function [168]. The first derivation of the MCSCF linear response function was accomplished in the late seventies [169, 170]. In 1985, Olsen and Jørgensen presented a general formulation of the MCSCF linear, quadratic, and cubic response functions [160]. Other MCSCF response methods have been given by Funchs et al. [171] and Kotuku et al. [172]. Second order perturbation theory response methods have been developed for linear response functions [173]. A formulation of time-dependent perturbation theory determining the frequency-dependent polarizability and first hyperpolarizability has been derived in the context of second order many-body perturbation theory (MP2) [174–177]. The latter approach has a pole structure determined by the SCF response equations and therefore uncorrelated (random phase approximation (RPA) poles). Coupled cluster response methods were initially derived by Monkhorst [178] and Dalgaard and Monkhorst [179], and a decade later a general formulation of coupled cluster linear and quadratic response functions was developed [180].

In general there is a fundamental difference between the second order many-body perturbation theory approach (MP2) and the other approaches. The MP2 approach involves two steps:

(1) The time development of an uncorrelated SCF wave function
(2) The utilization of the pole structure of the frequency dependent molecular properties of the SCF wave function to evaluate frequency dependent correlated MP2 molecular properties

The pole structure of the frequency-dependent molecular properties therefore occurs at the uncorrelated poles of the SCF wave function. Another disadvantage to this approach is that transition matrix elements, such as one- and two-photon transition moments, cannot be obtained in this approach. In all the other approaches, time development is determined for the correlated wave function and the pole structure is therefore also determined by the correlated wave function. Residues at the poles therefore give correlated transition matrix elements.

The above exposition basically describes the level of electronic structure response methods for calculating molecular properties of molecules in the gas phase.

15.4   A REACTION FIELD RESPONSE METHOD

Presently, the only response methods for calculating time-dependent molecular properties of solvated molecules can be found in refs. [133, 135, 180, 181]. The physical model is one where the solue molecule is embedded (contained in a spherical cavity) in a dielectric medium. The solute's charge distribution induces polarization moments in the dielectric medium. The polarization of the dielectric medium is described by the polarization vector relation

$$\mathbf{P}^{tot} = \mathbf{P}^{in} + \mathbf{P}^{op} \tag{15.35}$$

where the total polarization vector is given by $\mathbf{P}^{tot}$. The total polarization vector is the vector sum of the inertial polarization vector, $\mathbf{P}^{in}$ and the optical polarization vector $\mathbf{P}^{op}$. The total and optical polarization vectors of the dielectric medium are related to the static dielectric constant (ϵ_{st}) and optical dielectric constants (ϵ_{op}), respectively.

The polarization corresponding to the optical polarization vector stems from the response of the electronic degrees of freedom of the solvent or polymer, and it is assumed that the optical polarization vector responds instantaneously to changes in the solute charge distribution within the cavity. On the other hand the inertial polarization vector is related to the vibrational and molecular motion of the condensed phase, and this polarization has a much longer relaxation time compared to that of the optical polarization vector. The occurrence of sudden changes in the charge distribution of the solute within the cavity leads to a nonequilibrium solvation state with respect to the inertial polarization vector. Changes in the inertial polarization vector do not occur instantaneously, but with a relaxation time constant that is characteristic of the solvent. According to Eq. (15.35) the inertial polarization vector is obtained as the vector difference between the total and the optical polarization vectors.

The perturbation (the electric field in the case of electric polarizabilities) that is applied to the solvated/embedded molecule can either be time-independent or time-dependent. In the case of a time-dependent field, the frequencies of the externally applied field are much larger than the frequencies associated with the molecular vibrations. The electronic response of the solute molecule is assumed to be slower than the optical polarization and faster than the relaxation time of the inertial polarization. We assume that the optical polarization vector is always in equilibrium with the charge distribution of the solute.

The solute and solvent system ends up in a nonequilibrium state due to the high-frequency perturbation. The nonequilibrium state is represented by an optical polarization vector in equilibrium with the charge distribution of the solute, which is adjusted to the perturbation and an inertial polarization given by the charge distribution of the solute before the external field was turned on. The nonequilibrium state arises since the inertial polarization is not in equilibrium with the actual solute charge distribution, but instead to the charge distribution of the solute before application of the perturbation (see Chapter 14 for details). The nonequilibrium energy functional is given by [156, 158]

$$E_{\text{sol}} = \sum_{lm} R_{lm}(\rho, \epsilon_{\text{op}}) \langle T_{lm}(\rho) \rangle + R_{lm}(\rho', \epsilon_{\text{st}}, \epsilon_{\text{op}})(2 \langle T_{lm}(\rho) \rangle - \langle T_{lm}(\rho') \rangle)$$

(15.36)

where the total charge distribution of the solute after (before) the application of the perturbation is given by ρ (ρ'). The solute charge distribution is given by the expectation values of the charge moment operators T_{lm}. The response of the dielectric medium is given by the expectation values of the solvent polarization operators R_{lm}, which are related to the charge distributions of the solute. In the case of a solute contained within a spherical cavity having the radius a, they are given by the following relations

$$R_{lm}(\rho, \epsilon_{\text{op}}) = g_l(\epsilon_{\text{op}}) \langle T_{lm}(\rho) \rangle$$

(15.37)

and

$$R_{lm}(\rho', \epsilon_{\text{st}}, \epsilon_{\text{op}}) = g_l(\epsilon_{\text{st}}, \epsilon_{\text{op}}) \langle T_{lm}(\rho') \rangle$$

(15.38)

where the factors $g_l(\epsilon_{\text{op}})$ and $g_l(\epsilon_{\text{st}}, \epsilon_{\text{op}})$ are given by

$$g_l(\epsilon) = \frac{1}{2} \, a^{-(2l+1)} \, \frac{(l+1)(\epsilon - 1)}{1 + \epsilon(l+1)} \tag{15.39}$$

and

$$g_l(\epsilon_{st}, \epsilon_{op}) = g_l(\epsilon_{st}) - g_l(\epsilon_{op}) \tag{15.40}$$

We have performed a multipole expansion of the solute's charge distribution utilizing the procedure outlined in the previous chapter. The charge moments are given by the $\langle T_{lm}(\rho)\rangle$ elements and are expressed as expectation values of the nuclear and electronic solvent operators

$$\langle T_{lm}\rangle = T_{lm}^n - \langle T_{lm}^e\rangle \tag{15.41}$$

$$T_{lm}^e = t^{lm}(r) = \sum_{pq} t_{pq}^{lm} E_{pq} \tag{15.42}$$

$$T_{lm}^n = \sum_b Z_b t^{lm}(R_b) \tag{15.43}$$

where Z_b (R_b) is the nuclear charge (position vector) of the nucleus b. The subscripts p and q represent the orbitals ϕ_p and ϕ_q, and the operator E_{pq} is

$$E_{pq} = \sum_\sigma a_{p\sigma}^\dagger a_{q\sigma} \tag{15.44}$$

where we sum over the spin quantum number σ. The creation and annihilation operators for an electron in spin orbital $\phi_{p\sigma}$ are denoted $a_{p\sigma}^\dagger$ and $a_{p\sigma}$. We make use of the following notation

$$\langle T_{lm}^e\rangle = \frac{\langle 0|T_{lm}^e|0\rangle}{\langle 0|0\rangle} = \sum_{pq} D_{pq} t_{pq}^{lm} \tag{15.45}$$

where $|0\rangle$ is a reference electronic wave function for the solvated system. The elements t_{pq}^{lm} are given by

$$t_{pq}^{lm} = \langle \phi_p|t^{lm}|\phi_q\rangle \tag{15.46}$$

where D_{pq} is the one-electron density matrix. The functions $t^{lm}(Rb)$ and t_{pq}^{lm} are defined through the conventional spherical harmonics, S_l^μ, where μ is a positive integer [111, 112]

$$t^{l0} = S_l^0$$
$$t^{l\mu} = (2)^{-(1/2)}(S_l^\mu + S_l^{-\mu})$$
$$t^{l-\mu} = (-2)^{-(1/2)}(S_l^\mu - S_l^{-\mu}) \tag{15.47}$$

We let the electronic wave function of the solute be given by a (multiconfig-uration) self-consistent reaction field wave function $|0\rangle$. This reference wave function could correspond to the ground state or an excited state of the solvated system. The reference state is obtained by applying the variation principle to the total energy, E_{total}

$$E_{total} = E_{vac} + E_{sol} \tag{15.48}$$

where E_{vac} is the standard electronic energy for isolated molecules in a vac-uum and E_{sol} is the free energy of solvation (see Chapter 14: Nonequilibrium Solvation).

Let us consider the mathematical implication of introducing the solvent–solute interaction term into the Ehrenfest equation of motion of an operator B

$$\frac{d\langle B\rangle}{dt} = \left\langle\frac{\partial B}{\partial t}\right\rangle - i\langle[B, H]\rangle \tag{15.49}$$

where

$$H = H_0 + W_{sol} + V_{pert}(t) \tag{15.50}$$

The expectation values are determined from the time-dependent wave function $|O^t\rangle$ at time t

$$\langle\ldots\rangle = \langle{}^tO|\ldots|O^t\rangle \tag{15.51}$$

As in the previous chapter we assume that our solvated molecular system is described by a multiconfiguration self-consistent reaction field (MCSCRF) wave function. This wave function is fully optimized with respect to all orbital and configurational parameters and it satisfies the generalized Brillouin theorem

$$\begin{pmatrix} \langle O|[q^+, H_0 + W_{sol}]|O\rangle \\ \langle O|[R^+, H_0 + W_{sol}]|O\rangle \end{pmatrix} = \begin{pmatrix} 0 \\ 0 \end{pmatrix} \tag{15.52}$$

The wave function of the MCSCRF state is written as

$$|O\rangle = \sum_f C_f |\Phi_f\rangle \tag{15.53}$$

where the summation is over a set of configuration state functions (CSF) ($|\Phi_f\rangle$). Each configuration state function is a linear combination of Slater determinants

$$|\Phi_f\rangle = \prod_r a_r^\dagger |\text{vac}\rangle \tag{15.54}$$

where we make use of the standard ordered product of spin orbitals. We have used that

$$q_{pq}^\dagger = a_p^\dagger a_q \qquad p > q \tag{15.55}$$
$$R_n^\dagger = |n\rangle \langle O| \tag{15.56}$$

where $|n\rangle$ denotes the orthogonal complement space to $|O\rangle$.

The solvent–solute interaction part, W_{sol}, is given by

$$W_{\text{sol}} = \sum_{lm} g_l(\epsilon_{\text{op}})(T_{lm}^n)^2$$

$$+ \sum_{lm} g(\epsilon_{\text{st}}, \epsilon_{\text{op}})(T_{lm}^n)^2 + T_g(\epsilon_{\text{st}}, \epsilon_{\text{op}}) \tag{15.57}$$

The operator $T_g(\epsilon_{\text{st}}, \epsilon_{\text{op}})$ is made up of two very similar contributions, $T_1^g(\epsilon_{\text{op}})$ and $T_2^g(\epsilon_{\text{st}}, \epsilon_{\text{op}})$, which take the form

$$T_1^g(\epsilon_{\text{op}}) = \sum_{pq} \left(-2 \sum_{lm} g_l(\epsilon_{\text{op}}) \langle T_{lm}(\rho_f) \rangle t_{pq}^{lm} E_{pq} \right) \tag{15.58}$$

$$T_2^g(\epsilon_{\text{st}}, \epsilon_{\text{op}}) = \sum_{pq} \left(-2 \sum_{lm} g_l(\epsilon_{\text{st}}, \epsilon_{\text{op}}) \langle T_{lm}(\rho_i) \rangle t_{pq}^{lm} E_{pq} \right) \tag{15.59}$$

The expectation value of this effective solvent–solute interaction operator with respect to a state $|O\rangle$ is

$$\langle W_{\text{sol}} \rangle = \sum_{lm} g_l(\epsilon_{\text{op}})(T^n_{lm})^2 + \sum_{lm} g(\epsilon_{\text{st}}, \epsilon_{\text{op}})(T^n_{lm})^2 + \langle T_g(\epsilon_{\text{op}}) \rangle + \langle T_g(\epsilon_{\text{st}}, \epsilon_{\text{op}}) \rangle$$

$$= \sum_{lm} g_l(\epsilon_{\text{op}})\{(T^n_{lm})^2 - 2 \langle T_{lm}(\rho_f) \rangle \langle T^n_{lm}(\rho_f) \rangle\} + \sum_{lm} g_l(\epsilon_{\text{st}}, \epsilon_{\text{op}})(T^n_{lm})^2$$

$$- 2 \sum_{lm} g_l(\epsilon_{\text{st}}, \epsilon_{\text{op}}) \langle T_{lm}(\rho_i) \rangle \langle T^e_{lm}(\rho_f) \rangle$$

$$= \sum_{lm} g_l(\epsilon_{\text{op}})(\langle T_{lm}(\rho_f) \rangle^2 + \langle T^e_{lm}(\rho_f) \rangle^2)$$

$$+ \sum_{lm} g_l(\epsilon_{\text{st}}, \epsilon_{\text{op}})(2 \langle T_{lm}(\rho_f) \rangle \langle T_{lm}(\rho_i) \rangle + (T^n_{lm})^2 - 2 \langle T_{lm}(\rho_i) \rangle T^n_{lm})$$

$$= \sum_{lm} g_l(\epsilon_{\text{op}})(\langle T_{lm}(\rho_f) \rangle^2 + \langle T^e_{lm}(\rho_f) \rangle^2$$

$$+ \sum_{lm} g_l(\epsilon_{\text{st}}, \epsilon_{\text{op}})(2 \langle T_{lm}(\rho_f) \rangle \langle T_{lm}(\rho_i) \rangle - \langle T_{lm}(\rho_i) \rangle \langle T_{lm}(\rho_i) \rangle + \langle T_{lm}(\rho_i) \rangle^2)$$

$$= E_{\text{sol}} + \sum_{lm} g_l(\epsilon_{\text{op}}) \langle T^e_{lm}(\rho_f) \rangle^2 + \sum_{lm} g_l(\epsilon_{\text{st}}, \epsilon_{\text{op}}) \langle T^e_{lm}(\rho_i) \rangle^2 \qquad (15.60)$$

where E_{sol} is the nonequilibrium solvation energy. This illustrates that expectation values of W_{sol} are not the solvent contribution to the total energy [117, 135, 150], but are instead those corresponding to the effective solvent Hamiltonian satisfying the generalized Brillouin condition.

The Hamiltonian H_0 for the many-electronic system in a vacuum and the perturbation interaction operator representing the applied field are given as in Eq. (15.3).

The molecular system evolves in time after the application of the field and the time evolution of the molecular system may be parameterized as [160]

$$|O'\rangle = \exp(i\kappa(t)) \, \exp(iU(t))|O\rangle \qquad (15.61)$$

where $\exp(i\kappa(t))$ describes a unitary transformation in orbital space

$$\kappa = \sum_{rs} (\kappa_{rs}(t)a^{\dagger}_r a_s + \kappa'_{rs}(t)a^{\dagger}_s a_r) \qquad (15.62)$$

$$= \sum_{k} (\kappa_k(t)q^{\dagger}_k + \kappa'_k(t)q_k) \qquad (15.63)$$

and where the operator $\exp(iU(t))$ describes unitary transformations in configuration space

$$U(t) = \sum_n (S_n(t)R_n^{\dagger} + S_n^*(t)R_n) \qquad (15.64)$$

The evolution of the molecular system is described by the following set of operators

$$T = q_i^{\dagger}, R_i^{\dagger}, q_i, R_i \qquad (15.65)$$

These operators are collected in a row vector and a general vector in this basis is written as a column vector

$$\mathbf{N} = \begin{pmatrix} \kappa_j \\ S_j \\ \kappa_j' \\ S_j' \end{pmatrix} \qquad (15.66)$$

Thus N_j may refer to an orbital rotation parameter or a configuration parameter. The time evolution of $|O^t\rangle$ is determined subject to the requirement that Ehrenfest's theorem be satisfied through each order in the interaction operator.

We will now consider modifications arising from the solvent–solute interaction terms in the linear transformation with the $E^{[2]}$-matrix. Clearly the linear response function for an MCSCRF state has the same structure as the vacuum linear response function, the only difference being that some terms are added to the Hessian type matrix $E^{[2]}$ when including the outer medium. The $E^{[2]}$ matrix enters the linear response calculations either through the solution of sets of linear response equations, or through the linear response eigenvalue equations. In both cases, these equations are solved iteratively, where only linear transformations of $E^{[2]}$ on a trial vector $\mathbf{N}$ are carried out [160, 182–184]

$$d = E^{[2]}\mathbf{N} \qquad (15.67)$$

15.4.1   Ehrenfest's Equation of Motion

The operators to be considered in Ehrenfest's equation of motion are the time-transformed operators $\mathbf{T}^{\dagger,t}$ defined as

$$\mathbf{T}^{\dagger,t} = \begin{pmatrix} q^t \\ R^t \\ q^{\dagger,t} \\ R^{\dagger,t} \end{pmatrix} \qquad (15.68)$$

where the individual operators are given as

$$q_k^t = \exp(i\kappa(t)) \, q_k \, \exp(-i\kappa(t)) \tag{15.69}$$

$$q_k^{\dagger,t} = \exp(i\kappa(t)) \, q_k^\dagger \, \exp(-i\kappa(t)) \tag{15.70}$$

$$R_n^t = \exp(i\kappa(t)) \, \exp(iS(t)) \, R_n \, \exp(-iS(t)) \, \exp(-i\kappa(t)) \tag{15.71}$$

$$R_n^{\dagger,t} = \exp(i\kappa(t)) \, \exp(iS(t)) \, R_n^\dagger \, \exp(-iS(t)) \, \exp(-i\kappa(t)) \tag{15.72}$$

where $\kappa(t)$ and $S(t)$ are given in Eqs. (15.63) and (15.64). We obtain the response functions by solving Ehrenfest's equation for each order in the perturbation strength. When solving this equation we need to evaluate the expectation value of the commutator $[\mathbf{T}^{\dagger,t}, H]$, which can be written as

$$-i\langle[\mathbf{T}^{\dagger,t}, H]\rangle = -i\langle[\mathbf{T}^{\dagger,t}, H_0]\rangle - i\langle[\mathbf{T}^{\dagger,t}, V_{\text{pert}}(t)]\rangle - i\langle[\mathbf{T}^{\dagger,t}, W_{\text{sol}}]\rangle \tag{15.73}$$

of which the first two terms relate to the vacuum situation. For further details we refer to ref. [160].

We must consider the terms arising from $-i\langle[\mathbf{T}^{\dagger,t}, W_{\text{sol}}]\rangle$, which can be written as

$$-i\langle[\mathbf{T}^{\dagger,t}, W_{\text{sol}}]\rangle = -i\langle[\mathbf{T}^{\dagger,t}, T_1^g(\epsilon_{\text{op}})]\rangle - i\langle[\mathbf{T}^{\dagger,t}, T_2^g(\epsilon_{\text{st}}, \epsilon_{\text{op}})]\rangle \tag{15.74}$$

Thus the commutator involving W_{sol} gives rise to two terms containing the operators defined in Eqs. (15.58) and (15.59). Inserting the definitions of the operators $T_1^g(\epsilon_{\text{op}})$ and $T_2^g(\epsilon_{\text{st}}, \epsilon_{\text{op}})$ we obtain

$$
\begin{aligned}
&-i\langle[\mathbf{T}^{\dagger,t}, W_{\text{sol}}]\rangle \\[2mm]
&= -i\left\langle {}^t0\left|\left[\mathbf{T}^{\dagger,t}, \left\{2\sum_{lm,pq} g_l(\epsilon_{\text{op}}) t_{pq}^{lm} E_{pq}\right.\right.\right.\right. \\[2mm]
&\qquad\qquad \left.\left.\left.\left.\left(\sum_{p'q'} t_{p'q'}^{lm}\langle{}^t0|E_{p'q'}|0^t\rangle - T_{lm}^n\right)\right\}\right]\right|0^t\right\rangle \\[2mm]
&\quad -i\left\langle {}^t0\left|\left[\mathbf{T}^{\dagger,t}, \left\{-2\sum_{lm,pq} g_l(\epsilon_{\text{st}}, \epsilon_{\text{op}})\langle T_{lm}(\rho_i)\rangle t_{pq}^{lm} E_{pq}\right\}\right]\right|0^t\right\rangle
\end{aligned}
\tag{15.75}
$$

which may be written in terms of three contributions. Thus we have

$$-i\left\langle[\mathbf{T}^{\dagger,t}, W_{\text{sol}}]\right\rangle = G_{\text{sol}}^{(a)}(\mathbf{T}^{\dagger,t}) + G_{\text{sol}}^{(b)}(\mathbf{T}^{\dagger,t}) + G_{\text{sol}}^{(c)}(\mathbf{T}^{\dagger,t}) \tag{15.76}$$

with the following definitions:

$$G_{\text{sol}}^{(a)}(\mathbf{T}^{\dagger,t}) = -i\left\langle {}^t 0 \left| \left[\mathbf{T}^{\dagger,t}, \left\{ -2\sum_{lm,pq} g_l(\epsilon_{\text{op}}) t_{pq}^{lm} T_{lm}^n E_{pq} \right\} \right] \right| 0^t \right\rangle \tag{15.77}$$

$$G_{\text{sol}}^{(b)}(\mathbf{T}^{\dagger,t}) = -i\left\langle {}^t 0 \left| \left[\mathbf{T}^{\dagger,t}, \left\{ 2\sum_{lm,pq,p'q'} g_l(\epsilon_{\text{op}}) t_{pq}^{lm} t_{p'q'}^{lm} \langle {}^t 0 | E_{p'q'} | 0^t \rangle E_{pq} \right\} \right] \right| 0^t \right\rangle \tag{15.78}$$

$$G_{\text{sol}}^{(c)}(\mathbf{T}^{\dagger,t}) = -i\left\langle {}^t 0 \left| \left[\mathbf{T}^{\dagger,t}, \left\{ -2\sum_{lm,pq} g_l(\epsilon_{\text{st}}, \epsilon_{\text{op}}) \langle T_{lm}(\rho_i) \rangle t_{pq}^{lm} E_{pq} \right\} \right] \right| 0^t \right\rangle \tag{15.79}$$

The following section illustrates how to proceed in the case of the $G_{\text{sol}}^{(c)}(\mathbf{T}^{\dagger,t})$.

15.4.2 Contribution of Nonequilibrium Solvation Terms to Linear Response Equations

For simplicity we focus on the term in the response equations that relates to the inertial polarization response. We start out by introducing the following states:

$$|0^R\rangle = -\sum_n S_n R_n^\dagger |0\rangle = -\sum_n S_n |n\rangle \tag{15.80}$$

and

$$\langle 0^L| = \sum_n \langle 0|(S_n' R_n^\dagger) = \sum_n S_n' \langle n| \tag{15.81}$$

We also introduce the one-index-transformed solvent integrals

$$\sum_{pq} Q_{pq}^{lm} E_{pq} = \sum_{pq} [\kappa(t), t_{pq}^{lm} E_{pq}] \tag{15.82}$$

where

$$Q_{pq}^{lm} = \sum_{s} [\kappa_{ps} t_{sq}^{lm} - t_{ps}^{lm} \kappa_{sq}] \tag{15.83}$$

Consideration of $G_{\text{sol}}^{(c)}$ for the q_k^t component of the vector $\mathbf{T}^{\dagger,t}$ leads to

$$G_{\text{sol}}^{(c)}(q_k) = -i \left\langle 0 \left| \exp(-iS(t)) \exp(-i\kappa(t)) \left[\exp(i\kappa(t)) q_k \exp(-i\kappa(t)), \right. \right. \right.$$

$$\left. \left. \left\{ -2 \sum_{lm, pq} g_l(\epsilon_{\text{st}}, \epsilon_{\text{op}}) \langle T_{lm}(\rho_i) \rangle t_{pq}^{lm} E_{pq} \right\} \right] \right.$$

$$\left. \exp(i\kappa(t)) \exp(iS(t)) \right| 0 \right\rangle \tag{15.84}$$

Linear response calculations require terms linear in either $\kappa(t)$ or $S(t)$. Expanding the exponentials yields the following first order contribution

$$G_{\text{sol}}^{(c,1)}(q_k^t) = 2 \sum_{lm, pq} g_l(\epsilon_{\text{st}}, \epsilon_{\text{op}}) \langle T_{lm}(\rho_i) \rangle t_{pq}^{lm}$$

$$\{\langle 0 | [S(t), [q_k, E_{pq}]] | 0 \rangle + \langle 0 | [q_k, [\kappa(t), E_{pq}]] | 0 \rangle\} \tag{15.85}$$

For the operator $q_k^{\dagger}$ we obtain a similar expression by exchanging q_k with $q_k^{\dagger}$. In the case of the configurational operator R_n^t we obtain

$$G_{\text{sol}}^{(c,1)}(R_n^t) = 2 \sum_{lm, pq} g_l(\epsilon_{\text{st}}, \epsilon_{\text{op}}) \langle T_{lm}(\rho_i) \rangle t_{pq}^{lm}$$

$$\{\langle 0 | [R_n, [S(t), E_{pq}]] | 0 \rangle + \langle 0 | [R_n, [\kappa(t), E_{pq}]] | 0 \rangle\} \tag{15.86}$$

and in the case of $R_n^{\dagger}$ we obtain the same expression as Eq. (15.86) but with R_n replaced by $R_n^{\dagger}$. In the case of q_k^t, inserting the definitions of the orbital rotation operator $\kappa(t)$ yields

$$G_{\text{sol}}^{(c,1)}(q_k^t) = 2 \sum_{lm,pq} g_l(\epsilon_{\text{op}},\epsilon_{\text{st}}) \langle T_{lm}(\rho_i)\rangle \{(\langle 0|[q_k,E_{pq}]|0^R\rangle + \langle 0^L|[q_k,E_{pq}]|0\rangle) t_{pq}^{lm}$$

$$+ Q_{pq}^{lm} \langle 0|[q_k,E_{pq}]|0\rangle\} \tag{15.87}$$

and an analogous expression results in the case of $q_k^{\dagger,t}$. For the operators R_n^t and $R_n^{\dagger,t}$ we obtain

$$G_{\text{sol}}^{(c,1)}(R_n^t) = 2 \sum_{lm,pq} g_l(\epsilon_{\text{op}},\epsilon_{\text{st}}) \langle T_{lm}(\rho_i)\rangle$$

$$\{(S_n(t) \langle 0|E_{pq}|0\rangle + \langle n|E_{pq}|0^R\rangle) t_{pq}^{lm} + Q_{pq}^{lm} \langle n|E_{pq}|0\rangle\} \tag{15.88}$$

and

$$G_{\text{sol}}^{(c,1)}(R_n^{\dagger,t}) = 2 \sum_{lm,pq} g_l(\epsilon_{\text{op}},\epsilon_{\text{st}}) \langle T_{lm}(\rho_i)\rangle\{(S_n'(t) \langle 0|E_{pq}|0\rangle - \langle 0^L|E_{pq}|n\rangle) t_{pq}^{lm}$$

$$- Q_{pq}^{lm} \langle 0|E_{pq}|n\rangle\} \tag{15.89}$$

The first order contributions due to $G_{\text{sol}}^{(a)}$ and $G_{\text{sol}}^{(b)}$ for the different operators are given as

$$G_{\text{sol}}^{(a,1)}(q_k^t) = 2 \sum_{lm,pq} g_l(\epsilon_{\text{op}}) T_{lm}^n \{(\langle 0|[q_k,E_{pq}]|0^R\rangle + \langle 0^L|[q_k,E_{pq}]|0\rangle) t_{pq}^{lm}$$

$$+ Q_{pq}^{lm} \langle 0|[q_k,E_{pq}]|0\rangle\} \tag{15.90}$$

where the equation for $q_k^{\dagger,t}$ is the same as Eq. (15.90) but with q_k^t replaced by $q_k^{\dagger,t}$. For the operators $R_n^t, R_n^{\dagger,t}$ we find

$$G_{\text{sol}}^{(a,1)}(R_n^t) = 2 \sum_{lm,pq} g_l(\epsilon_{\text{op}}) T_{lm}^n \{(S_n(t) \langle 0|E_{pq}|0\rangle + \langle n|E_{pq}|0^R\rangle) t_{pq}^{lm}$$

$$+ Q_{pq}^{lm} \langle n|E_{pq}|0\rangle\} \tag{15.91}$$

$$G_{\text{sol}}^{(a,1)}(R_n^{\dagger,t}) = 2 \sum_{lm,pq} g_l(\epsilon_{\text{op}}) T_{lm}^n \{(S_n'(t) \langle 0|E_{pq}|0\rangle - \langle 0^L|E_{pq}|n\rangle) t_{pq}^{lm}$$

$$- Q_{pq}^{lm} \langle 0|E_{pq}|n\rangle\} \tag{15.92}$$

And further we have for the $G_{\text{sol}}^{(b)}$ terms

$$G_{\text{sol}}^{(b,1)}(q_k^t) = -2 \sum_{pq,lm,p'q'} g_l(\epsilon_{\text{op}}) t_{p'q'}^{lm} \left\{ (\langle 0|[q_k, E_{pq}]|0^R\rangle + \langle 0^L|[q_k, E_{pq}]|0\rangle) t_{pq}^{lm} \right.$$

$$+ Q_{pq}^{lm} \langle 0|[q_k, E_{pq}]|0\rangle \Big\} \langle 0|E_{p'q'}|0\rangle$$

$$- 2 \sum_{pq,lm,p'q'} g_l(\epsilon_{\text{op}}) t_{pq}^{lm} \left\{ (\langle 0|E_{p'q'}|0^R\rangle + \langle 0^L|E_{p'q'}|0\rangle) t_{p'q'}^{lm} \right.$$

$$+ Q_{p'q'}^{lm} \langle 0|E_{p'q'}|0\rangle \Big\} \langle |[q_k, E_{pq}]|0\rangle \tag{15.93}$$

where the $G_{\text{sol}}^{(b,1)}(q_k^{\dagger,t})$ equation is the same as Eq. (15.93) but with q_k^t replaced by $q_k^{\dagger,t}$. For the R_n^t and $R_n^{\dagger,t}$ operators we obtain

$$G_{\text{sol}}^{(b,1)}(R_n^t) = -2 \sum_{pq,lm,p'q'} g_l(\epsilon_{\text{op}}) t_{p'q'}^{lm} \left\{ (S(t) \langle 0|E_{pq}|0\rangle + \langle n|E_{pq}|0^R\rangle) t_{pq}^{lm} \right.$$

$$+ Q_{pq}^{lm} \langle n|E_{pq}|0\rangle \Big\} \langle 0|E_{p'q'}|0\rangle$$

$$- 2 \sum_{pq,lm,p'q'} g_l(\epsilon_{\text{op}}) t_{pq}^{lm} \left\{ (\langle 0|E_{p'q'}|0^R\rangle + \langle 0^L|E_{p'q'}|0\rangle) t_{p'q'}^{lm} \right.$$

$$+ Q_{p'q'}^{lm} \langle 0|E_{p'q'}|0\rangle \Big\} \langle n|E_{pq}|0\rangle \tag{15.94}$$

$$G_{\text{sol}}^{(b,1)}(R_n^{\dagger,t}) = -2 \sum_{pq,lm,p'q'} g_l(\epsilon_{\text{op}}) t_{p'q'}^{lm} \left\{ (S'(t) \langle 0|E_{pq}|0\rangle + \langle 0^L|E_{pq}|n\rangle) t_{pq}^{lm} \right.$$

$$- Q_{pq}^{lm} \langle 0|E_{pq}|n\rangle \Big\} \langle 0|E_{p'q'}|0\rangle$$

$$- 2 \sum_{pq,lm,p'q'} g_l(\epsilon_{\text{op}}) t_{pq}^{lm} \left\{ (\langle 0|E_{p'q'}|0^R\rangle + \langle 0^L|E_{p'q'}|0\rangle) t_{p'q'}^{lm} \right.$$

$$+ Q_{p'q'}^{lm} \langle 0|E_{p'q'}|0\rangle \Big\} \langle 0|E_{pq}|n\rangle \tag{15.95}$$

The addition of the linear contributions from $G_{\text{sol}}^{(a)}$, $G_{\text{sol}}^{(b)}$, and $G_{\text{sol}}^{(c)}$ is done most efficiently by defining the following operators:

$$T^g = -2 \sum_{lm,pq} \{g_l(\epsilon_{\rm op})\langle T_{lm}\rangle + g_l(\epsilon_{\rm op},\epsilon_{\rm st})\langle T_{lm}(\rho_i)\rangle\}t_{pq}^{lm}E_{pq} \qquad (15.96)$$

$$T^{yo} = -2 \sum_{lm,pq} \{g_l(\epsilon_{\rm op})\langle T_{lm}\rangle + g_l(\epsilon_{\rm op},\epsilon_{\rm st})\langle T_{lm}(\rho_i)\rangle\}Q_{pq}^{lm}E_{pq} \qquad (15.97)$$

$$T^{xo} = 2 \sum_{lm,pq} g_l(\epsilon_{\rm op})\langle Q_{lm}\rangle t_{pq}^{lm}E_{pq} \qquad (15.98)$$

$$T^{xc} = 2 \sum_{lm,pq} g_l(\epsilon_{\rm op})\{\langle 0|T_{lm}^e|0^R\rangle + \langle 0^L|T_{lm}^e|0\rangle\}t_{pq}^{lm}E_{pq} \qquad (15.99)$$

Then, using these effective operators one may collect the $G_{\rm sol}^{(a,1)}$, $G_{\rm sol}^{(b,1)}$, and $G_{\rm sol}^{(c,1)}$ terms for the individual elements of the vector $\mathbf{T}^{\dagger,t}$ such that we end up with

$$-i\langle[q_k,W_{\rm sol}]\rangle = -(\langle 0^L|[q_k,T^g]|0\rangle + \langle 0|[q_k,T^g]|0^R\rangle$$
$$+ \langle 0|[q_k,T^{xc}|0]\rangle) - \langle 0|[q_k,T^{yo}+T^{xo}|0]\rangle \qquad (15.100)$$

$$-i\langle[q_k^\dagger,W_{\rm sol}]\rangle = -(\langle 0^L|[q_k^\dagger,T^g]|0\rangle + \langle 0|[q_k^\dagger,T^g]|0^R\rangle$$
$$+ \langle 0|[q_k^\dagger,T^{xc}|0]\rangle) - \langle 0|[q_k^\dagger,T^{yo}+T^{xo}|0]\rangle \qquad (15.101)$$

$$-i\langle[R_n,W_{\rm sol}]\rangle = -(\langle n|T^g|0^R\rangle + \langle 0|T^g|0\rangle S_n(t)$$
$$+ \langle n|T^{xc}|0\rangle) - \langle n|T^{yo}+T^{xo}|0\rangle \qquad (15.102)$$

$$-i\langle[R_n^\dagger,W_{\rm sol}]\rangle = -(\langle 0^L|T^g|n\rangle + \langle 0|T^g|0\rangle S_n'(t)$$
$$- \langle n|T^{xc}|0\rangle) + \langle n|T^{yo}+T^{xo}|0\rangle \qquad (15.103)$$

These equations, Eqs. (15.100–15.103), are, along with the definitions of the effective one-electron operators in Eqs. (15.96–15.99), the mathematical ramifications of introducing the solute–solvent interactions into the response formalism. The implementation of this method as shown in ref. [181, 184] enables MCSRCF response calculations of properties of a molecular system surrounded by a dielectric medium in an equilibrium or nonequilibrium solvent state.

15.5   NUMERICAL EXAMPLES

As an example let us consider the hyperpolarizability of H_2O in water and present the effects of a first solvation shell on the hyperpolarizability of H_2O. For further details we refer to ref. [185]. Tables 15.1 and 15.2 illustrate that the RPA calculations give a reasonable description of the hyperpolarizabilities of H_2O in a vacuum. In Tables 15.3 and 15.4 we observe how the hyperpolarizabilities decrease when an H_2O molecule is solvated in a dielectric medium possess-

TABLE 15.1. MCSCF Calculations of the Hyperpolarizabilities of a Water Molecule in Vacuum

Frequency (a.u.)	β_{zyy} (a.u.)	β_{zxx} (a.u.)	β_{zzz} (a.u.)	β_z (a.u.)
$\omega = 0$	−8.32	−5.08	−10.63	−14.42
$\omega = 0.0345$	−8.40	−5.20	−10.77	−14.62
$\omega = 0.0556$	−8.60	−5.54	−11.15	−15.18
$\omega = 0.0932$	−8.91	−6.11	−11.74	−16.05

a.u., atomic units.

TABLE 15.2. RPA Calculations of the Hyperpolarizabilities of a Water Molecule in Vacuum

Frequency (a.u.)	β_{zyy} (a.u.)	β_{zxx} (a.u.)	β_{zzz} (a.u.)	β_z (a.u.)
$\omega = 0$	−7.37	−2.90	−7.84	−10.86
$\omega = 0.0345$	−7.43	−2.95	−7.93	−10.99
$\omega = 0.0556$	−7.60	−3.11	−8.18	−11.34
$\omega = 0.0932$	−7.86	−3.36	−8.56	−11.87

a.u., atomic units.

TABLE 15.3. MCSCRF Calculations of the Hyperpolarizabilities of a Water Molecule Surrounded by a Dielectric Medium with the Dielectric Properties of Liquid Water

Frequency (a.u.)	β_{zyy} (a.u.)	β_{zxx} (a.u.)	β_{zzz} (a.u.)	β_z (a.u.)
$\omega = 0$	−10.69	−5.70	−10.63	−16.18
$\omega = 0.0345$	−10.78	−5.80	−10.72	−16.39
$\omega = 0.0556$	−11.04	−6.11	11.09	−16.95
$\omega = 0.0932$	−11.42	−6.61	−11.64	−17.80

a.u., atomic units.

TABLE 15.4. RPA Calculations of the Hyperpolarizabilities of a Water Molecule Surrounded by a Dielectric Medium with the Dielectric Properties of Liquid Water

Frequency (a.u.)	β_{zyy} (a.u.)	β_{zxx} (a.u.)	β_{zzz} (a.u.)	β_z (a.u.)
$\omega = 0$	−9.32	−2.85	−7.41	−11.74
$\omega = 0.0345$	−9.40	−2.89	−7.49	−11.87
$\omega = 0.0556$	−9.61	−3.01	−7.72	−12.20
$\omega = 0.0932$	−9.92	−3.19	−8.05	−12.70

a.u., atomic units.

TABLE 15.5. RPA Calculations of the Hyperpolarizabilities of a Water Molecule Surrounded by a First Solvation Shell

Frequency (a.u.)	β_{zyy} (a.u.)	β_{zxx} (a.u.)	β_{zzz} (a.u.)	β_z (a.u.)
$\omega = 0$	−0.91	7.76	2.16	5.40
$\omega = 0.0345$	−0.91	7.82	2.17	5.45
$\omega = 0.0556$	−0.91	7.98	2.19	5.56
$\omega = 0.0932$	−0.90	8.22	2.22	5.72

a.u., atomic units.

TABLE 15.6. RPA Calculations of the Hyperpolarizabilities of a Water Molecule Surrounded by a First Solvation Shell and a Dielectric Medium (dielectric constants of liquid water)

Frequency (a.u.)	β_{zyy} (a.u.)	β_{zxx} (a.u.)	β_{zzz} (a.u.)	β_z (a.u.)
$\omega = 0$	−2.20	13.42	10.24	15.52
$\omega = 0.0345$	−2.26	13.56	10.37	15.72
$\omega = 0.0556$	−2.44	13.95	10.72	16.27
$\omega = 0.0932$	−2.74	14.52	11.25	17.11

a.u., atomic units.

ing the dielectric properties of water. We utilize the differential shell approach introduced in ref. [185], where one performs calculations of the hyperpolarizabilities for four different cases:

(1) The solvation shell and the H_2O molecule
(2) The solvation shell
(3) The solvation shell and the H_2O molecule plus the dielectric medium
(4) The solvation shell and the dielectric medium

The solvation shell effect on the hyperpolarizability of the solvated H_2O molecule is then obtained by subtracting the hyperpolarizabilities from the calculations denoted 1 and 2. The combined effect of the solvation shell and the dielectric medium is determined by subtracting the values of the hyperpolarizabilities from the calculations termed 3 and 4.

It is clearly seen from Tables 15.5 and 15.6 that the first solvation shell and the subsequent dielectric medium have a pronounced effect on the hyperpolarizability of the solvated H_2O molecule. It also illustrates the importance of including solvation shells when considering solutions or liquids with well-defined solvation shells.

16

MAGNETIC PROPERTIES OF SOLVATED MOLECULES

16.1 INTRODUCTION

In the previous chapter we considered a method for obtaining electric properties of solvated molecules. This chapter presents a rigorous method for obtaining solvent effects on magnetic properties such as magnetization and shielding constants. The method is based on work on electronic Hamiltonians for origin-independent studies of magnetic properties [186–189] and on dielectric medium response methods for calculating molecular properties of solvated molecules [111, 112, 156, 158, 180].

Again we assume that the solvent can be represented as a homogeneous, isotropic, linear dielectric medium and that the solvated molecule is contained within a spherical cavity embedded in the dielectric medium. We follow the procedure from Chapter 14 on nonequilibrium solvation by performing a multipole expansion of the charge distribution of the solvated molecule. As seen previously, the multipoles of the solvated molecule induce a reaction field that interacts with the molecular charge distribution. Since the externally applied magnetic field is time-independent, and since we are concerned with the time-independent magnetic properties, we utilize the equilibrium expression for the solvation energy.

The energy functional for the solvated molecule, described by a multiconfigurational self-consistent field electronic wave function, is

$$E(\mathbf{B}, \mathbf{m}) = E_{\text{vac}}(\mathbf{B}, \mathbf{m}) + E_{\text{sol}}(\mathbf{B}, \mathbf{m}) \tag{16.1}$$

where $E_{\text{vac}}(\mathbf{B}, \mathbf{m})$ is the energy corresponding to a Hamiltonian operator for a

194

vacuum state, and $E_{sol}(\mathbf{B}, \mathbf{m})$ is the equilibrium solvation energy. Both energy terms depend on the magnetic field induction $\mathbf{B}$ and the nuclear magnetic moments $\mathbf{m}$. The term $E_{vac}(\mathbf{B}, \mathbf{m})$ depends both explicitly on the external field plus the nuclear magnetic moments and implicitly, through the use of magnetic field dependent orbitals, on London orbitals [190]. The London atomic orbital (LAO) for an atomic orbital, χ_μ, centered on nucleus K having position vector $\mathbf{R}_K$ is given by

$$\omega_\mu(\mathbf{r}_K; \mathbf{A}^e) = \exp(-i\mathbf{A}^e \cdot \mathbf{r})\chi_\mu(\mathbf{r}_K) \tag{16.2}$$

where $\mathbf{A}^e$ is the vector potential for the external magnetic field at the nucleus K. The term $E_{sol}(\mathbf{B}, \mathbf{m})$ depends implicitly on the magnetic fields since the energy term will be evaluated using the magnetic-field-dependent orbitals.

We will consider the following magnetic properties of interest:

(1) The magnetization ξ which is given by

$$\xi = -\left.\frac{\partial^2 E(\mathbf{B}, \mathbf{m})}{\partial^2 \mathbf{B}}\right|_{\mathbf{B}=0} \tag{16.3}$$

(2) The nuclear magnetic shielding constant $\sigma(K)$ which is given by

$$\sigma(K) = 1 + \left.\frac{\partial^2 E(\mathbf{B}, \mathbf{m})}{\partial \mathbf{B}\, \partial \mathbf{m}_K}\right|_{\mathbf{B}=\mathbf{m}_K=0} \tag{16.4}$$

where $\mathbf{m}_K$ is the nuclear magnetic moment of the K-nucleus.

In the next section we present the method and illustrate how to obtain the equations necessary for determining gauge-origin independent magnetic properties of solvated molecules.

16.2   METHOD

We start off by establishing the energy functional, depending explicitly on the vector potential $\mathbf{A}$ due to the magnetic induction field $\mathbf{B}$. The expression for the solvation energy will be that corresponding to having the solute solvated by equilibrated solvent configurations. The expectation value of the vacuum Hamiltonian is

$$E_{vac}(\mathbf{B}, \mathbf{m}) = \frac{\langle O|H|O\rangle}{\langle O|O\rangle} \tag{16.5}$$

where we have used the spin-independent Born-Oppenheimer electronic Hamiltonian

$$H = \sum_{r,s} h_{rs} E_{rs} + \frac{1}{2} \sum_{rs,tu} (rs|tu) e_{rs,tu} \qquad (16.6)$$

The conventional one- and two-electronic integrals are denoted h_{rs} and $(rs|tu)$. The one- and two-electron spin-free operators are written as:

(1) The one-electron spin-free operator

$$E_{rs} = \sum_{\sigma} a_{r\sigma}^{\dagger} a_{s\sigma} \qquad (16.7)$$

(2) The two-electron spin-free operator

$$e_{rs,tu} = E_{rs} E_{tu} - \delta_{st} E_{ru} \qquad (16.8)$$

These expressions are summed over the spin quantum number σ. The operators $a_{p\sigma}^{\dagger}$ and $a_{p\sigma}$ are the standard creation and annihilation operators for an electron in spin orbital $\phi_{p\sigma}$.

The vector potential appears in a one-electron Hamiltonian where the kinetic energy operator has been modified. The one-electron Hamiltonian is (in a.u.) given by

$$h(\mathbf{r}; \mathbf{A}) = \frac{1}{2} \pi_i^2 - \sum_{K} \frac{Z_K}{r_{Ki}} \qquad (16.9)$$

where we sum over all nuclei specified by the position vector $\mathbf{R}_K$ and the charge Z_K. The term r_{Ki} is the electron–nucleus distance and is given by

$$r_{Ki} = |\mathbf{r}_i - \mathbf{R}_K| \qquad (16.10)$$

where $\mathbf{r}_i$ is the position vector of the ith electron. The kinetic momentum operator is modified by the presence of the vector potential and for the ith electron is given by

$$\pi_i = -i\nabla_i + \mathbf{A}(\mathbf{r}_i) \qquad (16.11)$$

with the vector potential given as

$$A(r) = \frac{1}{2} \, \mathbf{B} \times \mathbf{r}_O + \alpha^2 \sum_K \frac{\mathbf{m}_K \times \mathbf{r}_K}{r_K^3} \tag{16.12}$$

Here, α is the fine structure constant. The two terms in Eq. (16.12) represent the external magnetic field and the field from the nuclear magnetic moments, respectively. The arbitrary gauge origin is denoted O; the magnetic properties calculated should be independent of this choice. The field-dependent molecular orbitals provide origin-independent molecular integrals.

As reference electronic wave function we use an MCSCF wave function

$$|\Psi\rangle = \exp(-i\kappa) \exp(-iS)|O\rangle \tag{16.13}$$

where

$$|O\rangle = \sum_i C_i |\Theta_i\rangle \tag{16.14}$$

which is parameterized as in Chapter 14: Nonequilibrium Solvation. The configuration state functions (CSFs) are represented as $|\Theta_i\rangle$ and each of the CSFs is a linear combination of Slater determinants. The two unitary transformation operators are given by:

(1) Unitary transformation in orbital space

$$\kappa = \sum_{rs} [\kappa_{rs} a_r^\dagger a_s + \kappa_{sr}^* a_s^\dagger a_r] \tag{16.15}$$

(2) Unitary transformation in configuration space

$$S = \sum_n [S_n R_n^\dagger + S_n^* R_n] \tag{16.16}$$

where

$$R_n = |n\rangle\langle O| \tag{16.17}$$

and $|n\rangle$ denotes the orthogonal complement space to $|O\rangle$.

The energy contribution due to the dielectric medium is given by the equilibrium expression

$$E_{\text{sol}}(\mathbf{B}, \mathbf{m}) = \sum_{lm} g_l(\langle T_{lm}\rangle)^2 \tag{16.18}$$

where the function g_l depends on

(1) The cavity form and size
(2) The dielectric constant of the medium
(3) The moment expansion parameter

In the case of a spherical cavity of radius R_{cav} embedded in a medium with dielectric constant ϵ, the lth moment has

$$g_l = -\frac{1}{2} R_{\text{cav}}^{-(2l+1)} \frac{(l+1)(\epsilon-1)}{l+\epsilon(l+1)} \tag{16.19}$$

The charge moments of the charge distribution of the solute are given by the elements $\langle T_{lm}(\rho)\rangle$. The charge moments are given by the expectation values of the nuclear and electronic solvent operators

$$\langle T_{lm}\rangle = T_{lm}^n - \langle T_{lm}^e\rangle \tag{16.20}$$

$$T_{lm}^n = \sum_K Z_K R^{lm}(\mathbf{R}_K) \tag{16.21}$$

$$T_{lm}^e = R^{lm}(r) = \sum_{pq} R_{pq}^{lm} E_{pq} \tag{16.22}$$

where Z_K and $\mathbf{R}_K$ are the nuclear charge and position vector of the nucleus K, respectively. The following notation is used:

$$\langle T_{lm}^e\rangle = \frac{\langle O|T_{lm}^e|O\rangle}{\langle O|O\rangle} = \sum_{pq} D_{pq} R_{pq}^{lm} \tag{16.23}$$

where $|\Psi\rangle$ is the reference electronic wave function for the system under consideration [see Eq. (16.13)],

$$R_{pq}^{lm} = \langle \phi_p|R^{lm}|\phi_q\rangle \tag{16.24}$$

and D_{pq} is the one-electron density matrix.

As in the previous chapters on solvent effects on molecular properties, we define the functions $R^{lm}(\mathbf{R}_g)$ and R^{lm} through conventional spherical harmonics, S_l^μ, [111, 112]

$$R^{l0} = S_l^0$$

$$R^{l\mu} = (2)^{-(1/2)}(S_l^\mu + S_l^{-\mu})$$

$$R^{l-\mu} = (-2)^{-(1/2)}(S_l^\mu - S_l^{-\mu}) \tag{16.25}$$

where μ is a positive integer.

The energy functional depends on:

(1) The variational parameters in orbital and configuration space for the electronic wave function
(2) The magnetic field **B**
(3) The nuclear magnetic moments $\mathbf{m}_K$

We let the parameters referring to the magnetic field and magnetic moments be denoted (a, b) while the variational parameters for the electronic wave function are denoted (λ).

We focus on how to take into account the effects arising from the surrounding solvent and consider the optimization procedure of the second term in Eq. (16.1). The procedure for handling the optimization and characterization of points on the surface defined by the electronic variational parameters and magnetic parameters for the energy term E_{vac} is given in [187–189].

The derivatives of the energy functional in terms of wave function parameters, (λ_i) and magnetic parameters (a, b) are given by

$$\frac{\partial E}{\partial \lambda_i} = \frac{\partial E_{\text{vac}}}{\partial \lambda_i} + \frac{\partial E_{\text{sol}}}{\partial \lambda_i} \tag{16.26}$$

and

$$\frac{\partial E}{\partial a} = \frac{\partial E_{\text{vac}}}{\partial a} + \frac{\partial E_{\text{sol}}}{\partial a} \tag{16.27}$$

First we consider the gradient of the solvent term with respect to variation of wave function parameters given by

$$\frac{\partial E_{\text{sol}}}{\partial \lambda_i} = -2 \sum_{lm} g_l \langle T_{lm} \rangle \frac{\partial \langle T_{lm}^e \rangle}{\partial \lambda_i}$$

$$= -2 \sum_{lm} g_l \langle T_{lm} \rangle \sum_{pq} R_{pq}^{lm} \frac{\partial D_{pq}}{\partial \lambda_i} = \sum_{pq} t_{pq} \frac{\partial D_{pq}}{\partial \lambda_i} \tag{16.28}$$

where we have defined an effective one-electron operator

$$t_{pq} = -2 \sum_{lm} g_l \langle T_{lm} \rangle R_{pq}^{lm} \tag{16.29}$$

Next we consider the gradient with respect to variation of the magnetic parameters and we find

$$\frac{\partial E_{\text{sol}}}{\partial a} = -2 \sum_{lm} g_l \langle T_{lm} \rangle \frac{\partial \langle T_{lm}^e \rangle}{\partial a}$$

$$= -2 \sum_{lm} g_l \langle T_{lm} \rangle \sum_{pq} D_{pq} \frac{\partial R_{pq}^{lm}}{\partial a} = \sum_{pq} t_{pq}^a D_{pq} \tag{16.30}$$

where we have defined the effective one-electron operator

$$t_{pq}^a = -2 \sum_{lm} g_l \langle T_{lm} \rangle \frac{\partial R_{pq}^{lm}}{\partial a} \tag{16.31}$$

The third point we consider is the response of the electronic wave function with respect to the variation of the magnetic parameters; this term is given by

$$\frac{\partial^2 E_{\text{sol}}}{\partial \lambda_i \partial a} = 2 \sum_{lm} g_l \langle T_{lm} \rangle \frac{\partial^2 \langle T_{lm} \rangle}{\partial \lambda_i \partial a} + 2 \sum_{lm} g_l \frac{\partial \langle T_{lm} \rangle}{\partial \lambda_i} \frac{\partial \langle T_{lm} \rangle}{\partial a}$$

$$= t_{pq}^a \frac{\partial D_{pq}}{\partial \lambda_i} \tag{16.32}$$

where $\partial \langle T_{lm} \rangle / \partial a$ is zero, since the density matrix is symmetric whereas the derivative integrals with respect to a are antisymmetric. Finally, we consider the second derivative of the solvent energy term with respect to the magnetic parameters and obtain

$$\frac{d^2 E_{\text{sol}}}{db\, da} = \frac{\partial^2 E_{\text{sol}}}{\partial b\, \partial a} + \sum_i \frac{\partial^2 E_{\text{sol}}}{\partial b\, \partial \lambda_i} \frac{\partial \lambda_i}{\partial a} \tag{16.33}$$

where the latter term is the wave function response to the variation of the magnetic parameters, the term we considered above. The first term leads to

$$\frac{\partial^2 E_{sol}}{\partial b\, \partial a} = 2 \sum_{lm} g_l \langle T_{lm} \rangle \frac{\partial^2 \langle T_{lm} \rangle}{\partial b\, \partial a}$$

$$+ 2 \sum_{lm} g_l \frac{\partial \langle T_{lm} \rangle}{\partial b} \frac{\partial \langle T_{lm} \rangle}{\partial a}$$

$$= D_{pq} t_{pq}^{ab} \tag{16.34}$$

where we have defined the effective one-electron operator to be

$$t_{pq}^{ab} = -2 \sum_{lm} g_l \langle T_{lm} \rangle \frac{\partial^2 R_{pq}^{lm}}{\partial b\, \partial a} \tag{16.35}$$

As before the term involving $\partial \langle T_{lm} \rangle / \partial a$ is zero.

Technically we need to evaluate the solvent integrals with respect to the London atomic orbitals and furthermore to obtain the magnetic field derivatives of the solvent integrals. A procedure for that can be found in ref. [191].

16.3   NUMERICAL EXAMPLES

We will next consider the solvent effects on the magnetizability of water, and the nuclear shielding constant of hydrogen and oxygen for a solvated water molecule. We present Hartree-Fock gauge origin-independent calculations of these properties through the application of different models (for further details see ref. [192]):

(1) The H_2O molecule in vacuum
(2) The H_2O molecule surrounded by a dielectric medium having the dielectric properties of water
(3) The H_2O molecule surrounded by a first solvation shell
(4) The H_2O molecule surrounded by a first solvation shell and subsequently a dielectric medium that has the dielectric properties of water

As seen in Table 16.1, the magnetizability of water is strongly influenced by the presence of a solvation shell, while it depends insignificantly on the dielectric medium. In Tables 16.2 and 16.3 we present the results from the isotropic nuclear shielding constants of hydrogen and oxygen for the different solvent models, and again we observe that the first solvation shell has a strong influence on the magnetic properties of the solvated water molecule.

TABLE 16.1. The Magnetizability of Water for the Four Models

	ξ
Gas phase	-234.1
Dielectric medium	-234.7
1st Solvation shell	-199.7
1st Solvation shell and dielectric medium	-199.8

The units are $10^{-30}\ JT^{-2}$

TABLE 16.2. Results From the Solvent Model Studies of the Hartree-Fock Gauge Origin Independent Calculations of the Isotropic Nuclear Shielding Constant of Hydrogen

Hydrogen	σ^H
Gas phase	30.91
Dielectric medium	29.96
1st Solvation shell	27.36
1st Solvation shell and dielectric medium	26.94

The units are ppm.

TABLE 16.3. Results From the Solvent Model Studies of the Hartree-Fock Gauge Origin Independent Calculations of the Isotropic Nuclear Shielding Constant of Oxygen

Oxygen	σ^O
Gas phase	336.6
Dielectric medium	346.0
1st Solvation shell	318.8
1st Solvation shell and dielectric medium	320.2

The units are ppm.

17

QUANTUM MODEL FOR ELECTRON TRANSFER

This chapter contains:

(1) A general, brief overview of quantum mechanical models for electron transfer in solution

(2) A molecular approach for investigating the time evolution of electron transfer reactions in solution

For a more complete overview of this field we refer to the book by G. C. Schatz and M. A. Ratner [193]. The molecular approach utilizes concepts and models presented in Chapters 12 through 15 of this book.

17.1   INTRODUCTION

The system of interest will be considered to consist of:

(1) A solvated donor
(2) A solvated acceptor
(3) A surrounding solvent

Usually the methods consider the formation of a supermolecule, an encounter complex containing the donor and acceptor molecules, surrounded by a solvent [194–200]. The solvent is usually represented as a dielectric continuum [201–205] or as a phonon bath [194, 206–209]. The degrees of freedom for the total system are given as:

(1) The electronic degrees of freedom, denoted by the coordinate set $\{r_i\}$ of the solvated donor and acceptor

(2) The nuclear degrees of freedom, denoted by the coordinate set $\{Q_j\}$ of the solvated donor and acceptor

(3) The solvent coordinates, denoted by the coordinate set $\{q_k\}$.

Typically the last set of coordinates is constructed from the Fourier transform of the polarization vectors [194, 206–209] or is taken to be the inertial and the optical polarization vectors (see Chapter 14).

Within the Born-Oppenheimer approximation we write the Hamiltonian for the total system as

$$H_{tot}(\{r_i\}, \{Q_j\}, \{q_k\}) = H_{elec}(\{r_i\}, \{Q_j\}, \{q_k\}) + T_{mol}(\{Q_j\}) + T_{sol}(\{q_k\}) \quad (17.1)$$

where:

- $H_{elec}(\{r_i\}, \{Q_j\}, \{q_k\})$ is the electronic Hamiltonian for a given configuration of $\{Q_j\}, \{q_k\}$
- $T_{mol}(\{Q_j\})$ is the kinetic energy operator for the nuclear degrees of freedom of the solvated donor and acceptor
- $T_{sol}(\{q_k\})$ is the kinetic energy operator for the solvent degrees of freedom

We let the time-dependent wave function for the total system be described by

$$\Phi(\{r_i\}, \{Q_j\}, \{q_k\}, t) = \sum_{r,s,u} \phi_r(\{r_i\}, \{Q_j\}, \{q_k\}) \xi_s(\{Q_j\}, t) \zeta_u(\{q_k\}, t) \quad (17.2)$$

where

(1) The electronic wave function for the rth electronic state is $\phi_r(\{r_i\}, \{Q_j\}, \{q_k\}, t)$.

(2) The term $\xi_s(\{Q_j\}, t)$ is the nuclear wave function for the Q coordinates.

(3) The term $\zeta_u(\{q_k\}, t)$ is the wave function for the solvent coordinates.

For simplicity and to be in accordance with conventional approaches having electron transfer models for molecular systems we limit ourselves to two time-independent electronic wave functions [194–200, 206, 210–213], which represent the initial state (having the transferable electron on the donor) and the final state (having the transferable electron on the acceptor). For the sake of clarity we disregard the solvent terms in the Hamiltonian. With these approximations, the wave function for the electron transfer reaction becomes

$$\Phi(\{r_i\}, \{Q_j\}, t) = \phi_{in}(\{r_i\}, \{Q_j\})\xi_{in}(\{Q_j\}, t) + \phi_{fi}(\{r_i\}, \{Q_j\})\xi_{fi}(\{Q_j\}, t) \quad (17.3)$$

where

- $\phi_{in}(\{r_i\}, \{Q_j\})\xi_{in}(\{Q_j\}, t)$ represents the initial state
- $\phi_{fi}(\{r_i\}, \{Q_j\})\xi_{fi}(\{Q_j\}, t)$ represents the final state

Employing this expression for the wave function in the time-dependent Schrödinger equation

$$i\hbar \frac{\partial \Phi(\{r_i\}, \{Q_j\}, t)}{\partial t} = [H_{elec}(\{r_i\}, \{Q_j\}) + T_{mol}(\{Q_j\})]\Phi(\{r_i\}, \{Q_j\}, t) \quad (17.4)$$

we obtain, assuming that the overlap between the two electronic wavefunctions is zero,

$$i\hbar \frac{\partial \xi_x(\{Q_j\}, t)}{\partial t} = [H_{elec,xx} + T_{mol,xx} + T_{mol}(\{Q_j\})]\xi_x(\{Q_j\}, t)$$

$$+ [H_{elec,xy} + T_{mol,xy}]\xi_y(\{Q_j\}, t) \quad (17.5)$$

where $x \neq y$ and x or y denote either "in" or "fi." The various terms in the equation are:

(1) The term defining the Born-Oppenheimer surface for the electronic state x

$$H_{elec,xx} = \langle \phi_x(\{r_i\}, \{Q_j\}) | H_{elec}(\{r_i\}, \{Q_j\}) | \phi_x(\{r_i\}, \{Q_j\}) \rangle \quad (17.6)$$

(2) The term determining the electronic coupling between the two electronic states x and y

$$H_{elec,xy} = \langle \phi_x(\{r_i\}, \{Q_j\}) | H_{elec}(\{r_i\}, \{Q_j\}) | \phi_y(\{r_i\}, \{Q_j\}) \rangle \quad (17.7)$$

(3) The term that gives the Born-Oppenheimer correction terms for the electronic state x and thereby takes account of the Born-Oppenheimer breakdown

$$T_{\text{mol},xx} = \langle \phi_x(\{r_i\}, \{Q_j\}) | \sum_{Q_j} \frac{\hbar^2}{2M_{Q_j}} \nabla^2_{Q_j} | \phi_x(\{r_i\}, \{Q_j\}) \rangle$$

$$- \langle \phi_x(\{r_i\}, \{Q_j\}) | \sum_{Q_j} \frac{\hbar^2}{M_{Q_j}} \nabla_{Q_j} | \phi_x(\{r_i\}, \{Q_j\}) \rangle \times \nabla_{Q_j}$$

$$(17.8)$$

(4) Another Born-Oppenheimer breakdown operator giving the nuclear kinetic energy operator coupling between the two electronic states x and y

$$T_{\text{mol},xy} = \langle \phi_x(\{r_i\}, \{Q_j\}) | \sum_{Q_j} \frac{\hbar^2}{2M_{Q_j}} \nabla^2_{Q_j} | \phi_y(\{r_i\}, \{Q_j\}) \rangle$$

$$- \langle \phi_x(\{r_i\}, \{Q_j\}) | \sum_{Q_j} \frac{\hbar^2}{M_{Q_j}} \nabla_{Q_j} | \phi_y(\{r_i\}, \{Q_j\}) \rangle \times \nabla_{Q_j}$$

$$(17.9)$$

In the previous equations we have utilized the notation $\langle \dots | \dots | \dots \rangle$ to denote integration over electronic coordinates. The mass M_{Q_j} is the effective mass for the coordinate Q_j.

The electron transfer between the donor and acceptor sites arises due to the second term in Eq. (17.5), and this term determines whether or not the reaction will occur.

The solutions $(\xi^0_{x,v}(\{Q_j\}))$ and eigenvalues $(E_{x,v})$ of the time-independent nuclear Schrödinger equation

$$[H_{\text{elec},xx} + T_{\text{mol},xx} + T_{\text{mol}}(\{Q_j\})]\xi^0_{x,v}(\{Q_j\}) = E_{x,v}\xi^0_{x,v}(\{Q_j\}) \qquad (17.10)$$

enable us to write the time-dependent nuclear wave functions as

$$\sum_v \alpha_{x,v}\xi^0_{x,v}(\{Q_j\}) \exp\left(-i\frac{E_{x,v}t}{\hbar}\right) \qquad (17.11)$$

which will be used in the following subsection.

The assumption of orthogonality of the electronic wave functions is easily circumvented as illustrated in [212, 214]. As electronic wave functions one uses:

(1) A set that does not diagonalize the electronic Hamiltonian. These states are denoted diabatic states. In this case the coupling between the two electronic states occurs through $H_{elec,xy}$ and the Born-Oppenheimer breakdown terms are assumed to be zero [211, 212, 215].

(2) A set that diagonalizes the electronic Hamiltonian. The resulting states are termed adiabatic states. Here, the coupling stems from $T_{mol,xy}$, and the term $H_{elec,xy}$ is zero by definition.

The diabatic representation leads to the following time-dependent equation of motion

$$i\hbar \frac{\partial \xi_x(\{Q_j\},t)}{\partial t} = [H_{elec,xx} + T_{mol}(\{Q_j\})]\xi_x(\{Q_j\},t) + H_{elec,xy}\xi_y(\{Q_j\},t)$$

(17.12)

and in the adiabatic representation we obtain

$$i\hbar \frac{\partial \xi_x(\{Q_j\},t)}{\partial t} = [H_{elec,xx} + T_{mol,xx} + T_{mol}(\{Q_j\})]\xi_x(\{Q_j\},t) + T_{mol,xy}\xi_y(\{Q_j\},t)$$

(17.13)

Next we consider an analytical expression for the rate constant of an electron transfer reaction in the quantum mechanical regime.

17.1.1 Rate Constant for Electron Transfer in the Quantum Mechanical Regime

The classical model of Marcus [123] is generally used when coupling between the donor and acceptor sites is large, and for cases where a classical description of the nuclear motion suffices. The quantum mechanical description is utilized when either the electronic coupling is small or nuclear tunneling is of importance. The equation of motion in Eq. (17.12) is solved by first order time-dependent perturbation theory and the expression for the electron transfer rate constant is [194–196, 206, 216–219]

$$k_{et} = \frac{2\pi}{\hbar} \frac{1}{q_{vib,in}} \sum_{w,v} \exp\left[-\frac{E_{in,v}}{k_B T}\right]$$

$$\cdot [(\xi^0_{in,v}(\{Q_j\})|\langle\phi_{in}(\{r_i\},\{Q_j\})|H_{elec}(\{r_i\},\{Q_j\})|\phi_{fi}(\{r_i\},\{Q_j\}))|\xi^0_{fi,w}(\{Q_j\}))]^2$$

$$\cdot \delta(E_{in,v} - E_{fi,w})$$

(17.14)

where

- $q_{\text{vib, in}}$ is the vibrational partition function for the initial state
- $E_{\text{in}, v}$ and $E_{\text{fi}, w}$ are the energies in the initial and final states, respectively
- k_B and T are the Boltzmann constant and the temperature, respectively

The electronic and nuclear wave functions are as defined in the previous section. Application of the Franck-Condon approximation leads to a factorization of the coupling element, and upon applying this approximation we obtain

$$(\xi_{\text{in}, v}^0(\{Q_j\}) | \langle \phi_{\text{in}}(\{r_i\}, \{Q_j\}) | H_{\text{elec}}(\{r_i\}, \{Q_j\}) | \phi_{\text{fi}}(\{r_i\}, \{Q_j\}) \rangle | \xi_{\text{fi}, w}^0(\{Q_j\}))$$

$$= (\xi_{\text{in}, v}\{Q_j\}) | \xi_{\text{fi}, w}^0(\{Q_j\})) \langle \phi_{\text{in}}(\{r_i\}, \{Q_j\}) | H_{\text{elec}}(\{r_i\}, \{Q_j\}) | \phi_{\text{fi}}(\{r_i\}, \{Q_j\}) \rangle$$

$$(17.15)$$

where the electronic coupling is evaluated at some appropriate nuclear configuration or as a function of the nuclear degrees of freedom.

17.2   THE MODEL HAMILTONIAN FOR THE SYSTEM

The purpose of this section is to outline a model Hamiltonian for a molecular and time-dependent description of electron transfer reactions in solution. The Hamiltonian is used, together with time-dependent statistical density operators and Ehrenfest's theorem, to derive a set of time-dependent equations of motion for electron transfer in solution. The model follows the electron transfer reaction between a donor and acceptor site within a molecular system enclosed by a spherical cavity. The spherical cavity is embedded in a dielectric medium.

The total Hamiltonian is written as a sum of five parts

$$H = H_o + H_p + W_{o-p} + H_v + W_{e-v} \qquad (17.16)$$

where

- H_o is the usual many-electron Hamiltonian for the molecular system in the cavity
- H_p is a model Hamiltonian for the solvent degrees of freedom
- W_{o-p} describes the dielectric interactions between the dielectric medium and the molecular system within the cavity
- H_v is the Hamiltonian for the intramolecular vibrational degrees of freedom for the molecular system in the cavity
- W_{e-v} describes the interactions between the electronic and intramolecular degrees of freedom for the molecular system within the cavity

The construction and characterization of the Hamiltonian will be performed in seven main steps.

17.2.1   The Statistical Density Operator

A time-dependent statistical density operator is used for describing the evolution of the total system. We assume that we can write it as a direct product

$$\rho = \rho_{molecule} \otimes \rho_{medium} = \rho_0 \otimes \rho_s \qquad (17.17)$$

where the subscripts o and s represent the molecular subsystem and the medium subsystem, respectively. The respective density operators are normalized

$$\text{Tr}_o\{\rho_o\} = \text{Tr}_s\{\rho_s\} = 1 \qquad (17.18)$$

which means

$$\text{Tr}\{\rho\} = 1 \qquad (17.19)$$

Having a molecular operator C and a medium operator D, we obtain from Eq. (17.17)

$$\langle C \rangle = \text{Tr}_o\{C\rho_0\} \qquad (17.20)$$
$$\langle D \rangle = \text{Tr}_s\{D\rho_s\} \qquad (17.21)$$
$$\langle CD \rangle = \langle C \rangle \langle D \rangle \qquad (17.22)$$

17.2.2   The Dielectric Medium

As in Chapter 14 we have to introduce the optical and inertial polarization and use the free energy functional for equilibrium and nonequilibrium solvent configurations. At present we are only concerned with abstract degrees of freedom for the outer medium.

The molecular charge distribution induces polarization charges in the dielectric medium. The option of inducing charge density configurations in the dielectric medium has its origin in the displacement of the degrees of freedom in the dielectric medium. At present we do not need to know precisely what these degrees of freedom are—only that they are some kind of generalized coordinates. We assume that the solvent degrees of freedom fulfill boson algebra. Using Eq. (17.17) we can write

$$\langle H_o \rangle = \text{Tr}\{\rho H_o\} = \text{Tr}\{\rho_o H_o\} = E_o \qquad (17.23)$$

The total energy change associated with the introduction of charges in the cavity is

$$\langle H - E_o - E_{o,p}\rangle = \frac{1}{2}\int_V d\mathbf{r}\rho(\mathbf{r})U(\mathbf{r})$$

$$= -\frac{1}{2}\sum_k f_k\langle M_k^\dagger\rangle\langle M_k\rangle \tag{17.24}$$

where

(1) The term $E_{o,p}$ is the energy of the above-mentioned bosons in the absence of charges in the cavity.

(2) The index k indicates the quantum numbers l and m.

(3) The term $\langle M_k\rangle$ is the expectation value of the charge moment operator M_k of the molecular charge distribution within the cavity

$$\langle M_k\rangle = \mathrm{Tr}\{\rho M_k\} = \mathrm{Tr}_o\{\rho_o M_k\} \tag{17.25}$$

From ref. [220] and Chapter 14 we have that

$$\langle W_{o-p}\rangle = \int \rho(\mathbf{r})U(\mathbf{r})d\mathbf{r} = -\sum_k f_k\langle M_k^\dagger\rangle\langle M_k\rangle \tag{17.26}$$

and the expectation value for H_p is given by

$$\langle H_p\rangle = \frac{1}{2}\int_S da \int_S da'\, \frac{\sigma_p(\bar{\mathbf{r}})\rho_p(\mathbf{r}')}{|\mathbf{r} - \mathbf{r}'|}$$

$$= E_{o,p} - \frac{1}{2}\int d\mathbf{r}\,\rho(\mathbf{r})U(\mathbf{r})$$

$$= E_{o,p} + \frac{1}{2}\sum_k f_k\langle M_k^\dagger\rangle\langle M_k\rangle \tag{17.27}$$

17.2.3 The Solvent Model Hamiltonian

From the structure of Eqs. (17.16–17.27) one notices similarities with the equation encountered for interactions between molecular system and quantum mechanical radiation fields [221]. For the molecular–solvent part we apply the model Hamiltonian

$$H' = H_o + H_p + W_{o-p} = H_o + \sum_k w_k b_k^\dagger b_k + \sum_k \{\alpha_k M_k b_k^\dagger + \alpha_k^* M_k^\dagger b_k\} \quad (17.28)$$

where the coupling between the two subsystems is given by the α_k-parameters. We compare Eq. (17.28) and Eq. (17.16), and it is easily seen that

$$H_p = \sum_k w_k b_k^\dagger b_k \quad (17.29)$$

and

$$W_{o-p} = \sum_k \{\alpha_k M_k b_k^\dagger + \alpha_k^* M_k^\dagger b_k\} \quad (17.30)$$

where the set of solvent operators $\{b_k, b_k^\dagger\}$ satisfy boson algebra.

17.2.4   The Statistical Density Operator for the Solvent

When the solvent is unperturbed by the presence of the molecular system it is assumed to be described by a statistical density operator ρ_s that corresponds to a Boltzmann distribution at temperature T:

$$\rho_s(0) = \frac{\exp[-\beta \sum_k w_k b_k^\dagger b_k]}{\text{Tr}\{\exp(-\beta \sum_k w_k b_k^\dagger b_k)\}} \quad (17.31)$$

and to have the following properties:

(1) The occupation number q_k is for the kth mode

$$q_k \equiv \text{Tr}_s\{\rho_s(0)b_k^\dagger b_k\} = [\exp(w_k \beta) - 1]^{-1} \quad (17.32)$$

(2) The expectation value of b_k is

$$\text{Tr}_s\{\rho_s(0)b_k\} = 0 \quad (17.33)$$

The interactions between the molecular system within the cavity and the dielectric medium change the statistical density operator ρ_s, which is obtained from the unperturbed medium density operator $\rho_s(0)$ by performing a unitary transformation. This unitary transformation is denoted a Glauber transformation [133].

$$\rho_s = \rho_s(X_1, X_1, \ldots, X_n, \ldots)$$
$$= \exp(iS)\,\rho_s(0)\exp(-iS) \qquad (17.34)$$

where

$$S = \sum_k (X_k b_k^\dagger + X_k^* b_k) \qquad (17.35)$$

Furthermore we have, using Eq. (17.34),

(1) The expectation value of b_k

$$\langle b_k \rangle = \mathrm{Tr}\{\rho b_k\} = \mathrm{Tr}_s\{\rho_s(0)\exp(-iS)\,b_k\exp(iS)\} = iX_k \qquad (17.36)$$

(2) The expectation value of $b_k^\dagger$

$$\langle b_k^\dagger \rangle = -iX_k^* \qquad (17.37)$$

(3) The expectation value of $b_k^\dagger b_k$

$$\langle b_k^\dagger b_k \rangle = q_k + X_k^* X_k \qquad (17.38)$$

17.2.5   Variational Selection of Initial State

Next we consider determining variationally the parameters $\{X_k\}$ of the Glauber transformation. Using Eq. (17.17), we obtain the expectation value of the parts in the Hamiltonian that refer to the molecular-solvent system

$$\langle H' \rangle = \langle H_o + H_p + W_{o-p} \rangle = E_o + \langle H_p + W_{o-p} \rangle \qquad (17.39)$$

and using Eqs. (17.28) and (17.34), and Eqs. (17.37) and (17.38), we find that

$$\langle H' \rangle = \mathrm{Tr}\{\rho_o \exp(iS)\,\rho_s(0)\exp(-iS)H'\}$$
$$= E_o + \sum_k w_k(q_k + X_k^* X_k)$$
$$+ \sum_k [-iX_k^* \alpha_k \langle M_k \rangle + iX_k \alpha_k^* \langle M_k^\dagger \rangle] \qquad (17.40)$$

We determine the variation of $\langle H' \rangle$ with respect to X_k^* and X_k and we find that

$$\frac{\partial}{\partial X_k^*} \langle H' \rangle = w_k X_k - i\alpha_k \langle M_k \rangle = 0 \qquad (17.41)$$

$$\frac{\partial}{\partial X_k} \langle H' \rangle = w_k X_k^* + i\alpha_k^* \langle M_k^\dagger \rangle = 0 \qquad (17.42)$$

We are thereby able to select the optimal parameters for the Glauber transformation and we find that

$$X_k = i\alpha_k w_k^{-1} \langle M_k \rangle \qquad (17.43)$$

$$X_k^* = -i\alpha_k^* w_k^{-1} \langle M_k^\dagger \rangle \qquad (17.44)$$

which means that we have selected an initial state for the system corresponding to a state of minimal energy. We have achieved this by variationally determining the values of the parameters for the Glauber transformation. The variational scheme for choosing the initial state is a more appropriate method for obtaining the initial state than the scheme used in the more conventional perturbative models. Perturbative models use an arbitrary equilibrium precursor configuration [194, 215].

17.2.6   The Hamiltonian for the Intramolecular Degrees of Freedom of the Molecular System

The molecular vibrations (the intramolecular degrees of freedom) of the molecular system within the cavity are described by

$$H_v = \sum_n \Omega_n B_n^\dagger B_n \qquad (17.45)$$

where:

(1) The frequency of the mode n is Ω_n.
(2) The boson creation (annihilation) operator for the mode n is $B_n^\dagger$ (B_n).

The interaction operator between the electronic subsystem and the molecular vibrations is obtained from a Taylor expansion of the full electronic Hamiltonian in the intramolecular degrees of freedom to first order around a given intramolecular configuration $\{Q_i^o\}$

$H_o(\{r_a\}, \{Q_b\}, \{k_c\})$

$\quad = H_o(\{r_a\}, \{Q_i^o\}, \{k_c\})$

$$+ \sum_n \left. \frac{\partial H_o(\{r_a\}, \{Q_b\}, \{k_c\})}{\partial Q_n} \right|_{Q_n = Q_n^o} (Q_n - Q_n^o)$$

$$+ \frac{1}{2} \sum_{n,n'} \left. \frac{\partial^2 H_o(\{r_a\}, \{Q_b\}, \{k_c\})}{\partial Q_n \partial Q_{n'}} \right|_{Q_n = Q_n^o, Q_{n'} = Q_{n'}^o} (Q_n - Q_n^o)(Q_{n'} - Q_{n'}^o)$$

$$+ \cdots \tag{17.46}$$

where

- $\{r_a\}$ is the set of electronic coordinates
- $\{Q_b\}$ is the set of intramolecular degrees of freedom
- $\{k_c\}$ is the set of degrees of freedom for the outer solvent

To first order we obtain

$$W_{e-v} = \sum_n A_n H_n^{(1)}(B_n^\dagger + B_n) \tag{17.47}$$

where we have introduced:

(1) The first order term

$$H_n^{(1)} = \left. \frac{\partial H_o(\{r_a\}, \{Q_b\}, \{k_c\})}{\partial Q_n} \right|_{Q_n = Q_n^o} \tag{17.48}$$

(2) The displacement operators

$$(Q_n - Q_n^o) = \sqrt{\frac{1}{2M_n \Omega_n}} \, (B_n^\dagger + B_n) \tag{17.49}$$

where $\Omega_n(M_n)$ is the frequency (reduced mass) of the mode n
(3) The parameter A_n

$$A_n = \sqrt{\frac{1}{2M_n \Omega_n}} \tag{17.50}$$

17.2.7   The Statistical Density Operator for the Molecular System

The statistical density operator for the molecular system within the cavity is given by

$$\rho_o = |\Phi\rangle\langle\Phi| \otimes \rho_v \qquad (17.51)$$

where

- $|\Phi\rangle$ is the electronic wave function for the molecular system
- ρ_v is the statistical density operator for the intramolecular degrees of freedom.

For the latter we utilize an approach similar to that used for the solvent degrees of freedom. When the intramolecular degrees of freedom are unperturbed by the coupling to the electronic system we have

$$\rho_v(0) = \frac{\exp[-\beta \sum_n \Omega_n B_n^\dagger B_n]}{\text{Tr}\{\exp[-\beta \sum_n \Omega_n B_n^\dagger B_n]\}} \qquad (17.52)$$

and this statistical density operator has the following properties:

(1) The occupation number for the nth mode is given by

$$\theta_n = \text{Tr}\{B_n^\dagger B_n \rho_v(0)\} = \frac{1}{\exp[\beta\Omega_n] - 1} \qquad (17.53)$$

(2) The expectation value of the operator B_n is

$$\langle B_n\rangle = \text{Tr}\{B_n \rho_v(0)\} \qquad (17.54)$$

As soon as we introduce the coupling between the electronic and intramolecular degrees of freedom we obtain the new statistical density operator by a unitary transformation

$$\rho_v = \exp(iV)\rho_v(0)\exp(-iV) \qquad (17.55)$$

where the operator V is represented by

$$V = \sum_n (Y_n B_n^\dagger + Y_n^* B_n) \qquad (17.56)$$

and for the expectation values we find that

- $\langle B_n^\dagger \rangle$ is given by

$$\langle B_n^\dagger \rangle = -i Y_n^* \tag{17.57}$$

- $\langle B_n \rangle$ is given by

$$\langle B_n \rangle = i Y_n \tag{17.58}$$

- $\langle B_n^\dagger B_n \rangle$ is given by

$$\langle B_n^\dagger B_n \rangle = \theta_n + Y_n^* Y_n \tag{17.59}$$

To select an appropriate initial state for ρ_v we can use the variational approach we used for the transformation parameters for the statistical density operator for the solvent. This gives us the option of having both intramolecular and solvent configurations corresponding to the correct lowest energy configuration for the initial state.

17.3   THE TIME-DEPENDENT EQUATION OF MOTION AND THE ELECTRON TRANSFER PROBABILITY

In the case of the time-dependent problem we again make the assumption that the density operator is factorized

$$\rho(t) = \rho_o(t) \otimes \rho_s(t) \tag{17.60}$$

where ρ_0 and ρ_s as before denote the molecular and solvent density operators, respectively. The time evolution is governed by Ehrenfest's theorem [222], [Eq. (17.61)], for arbitrary operators

$$i\hbar \frac{d}{dt} \langle \mathcal{A} \rangle = \langle [\mathcal{A}, H] \rangle \tag{17.61}$$

For the solvent density operators we employ the ansatz

$$\rho_s(t) = \exp(iS(t)) \rho_s(0) \exp(-iS(t)) \tag{17.62}$$

with

$$S(t) = \sum_k (X_k(t)b_k^\dagger + X_k^*(t)b_k) \tag{17.63}$$

where $X_k(0)$ is determined through the optimization of the Glauber transformation from the previous section. The molecular density operator is defined by

$$\rho_o(t) = |\Phi(t)\rangle\langle\Phi(t)| \tag{17.64}$$

and

$$|\Phi(t)\rangle = \alpha(t)|D^-,A\rangle + \beta(t)|D,A^-\rangle \tag{17.65}$$

where the initial state is given by $|D^-,A\rangle$ and the final state by $|D,A^-\rangle$. For the ET reaction the initial conditions are

$$\alpha(0) = 1 \qquad \beta(0) = 0 \tag{17.66}$$

In order to determine an expression for the ET probability we require that Eq. (17.61) hold for the operators

$$\{a_D^\dagger a_A, a_A^\dagger a_D, B_n^\dagger, B_n, b_k^\dagger, b_k\} \equiv \{q_i, q_i^\dagger, B_n^\dagger, B_n, b_k^\dagger, b_k\} \tag{17.67}$$

The equation of motion for these operators is given by

$$i\frac{d}{dt}\langle q_i\rangle = \langle[q_i,H]\rangle \tag{17.68}$$

$$i\frac{d}{dt}\langle B_n\rangle = \langle[B_n,H]\rangle \tag{17.69}$$

$$i\frac{d}{dt}\langle b_k\rangle = \langle[b_k,H]\rangle \tag{17.70}$$

Focussing on the electronic operators and the solvent operators we find by using Eqs. (17.60–17.65), together with Eqs. (17.68) and (17.70) that

$$i\frac{d}{dt}\langle q_i\rangle_o = \langle[q_i,\tilde{H}]\rangle \tag{17.71}$$

$$i\frac{d}{dt}\langle\tilde{b}_k\rangle_s = \langle[\tilde{b}_k,\tilde{H}]\rangle \tag{17.72}$$

where

$$\langle Q \rangle_o = \text{Tr}\{\rho_o(t)Q\} \tag{17.73}$$

$$\langle Q \rangle_s = \text{Tr}\{\rho_s(t)Q\} \tag{17.74}$$

$$\tilde{b}_k = \exp(-iS)b_k \exp(iS) \tag{17.75}$$

$$= b_k + iX_k(t)$$

$$\langle \tilde{b} \rangle = iX_k(t) \tag{17.76}$$

$$\tilde{H} = \exp(-iS)H \exp(iS)$$

$$= H_o + \sum_k w_k \tilde{b}_k^\dagger \tilde{b}_k + \sum_k \{\alpha_k M_k \tilde{b}_k^\dagger + \alpha_k^* M_k^\dagger \tilde{b}_k\} \tag{17.77}$$

After determining the commutator between the transformed field operators of the medium and the transformed Hamiltonian and using Eqs. (17.75) and (17.76), we obtain the time evolution of the Glauber transformation parameters

$$i \frac{d}{dt} X_k(t) = w_k X_k(t) - i\alpha_k \langle M_k \rangle_o \tag{17.78}$$

This equation has to be solved numerically along with

$$i \frac{d}{dt} \langle q_i \rangle_o = \langle [q_i, H_o] \rangle_o + \langle [q_i, \tilde{W}_{o-p}] \rangle \tag{17.79}$$

Furthermore we have to divide the solvent degrees of freedom into two classes corresponding to the optical and inertial polarization. Here we utilize our knowledge from Chapter 14 on nonequilibrium solvation. We let a class of $X_k(t)$ parameters follow the instantaneous changes occurring within the molecular system, while another class can:

(1) Remain fixed during the process
(2) Change according to the kinetic energy alloted to these degrees of freedom
(3) Reorient themselves according to the Debye relaxations times for the medium

In the next section we present a few numerical examples illustrating some of the features of the model.

17.4 NUMERICAL EXAMPLES

In Fig. 17.1 we present calculations that have utilized the nonequilibrium solvent energy expression presented in Chapter 14. The calculations are performed

Diabatic Free Energy Curves vs. Inertial Polarization.
Donor-acceptor distance is 4.0Å.
The intramolecular configuration is that of the reactants.

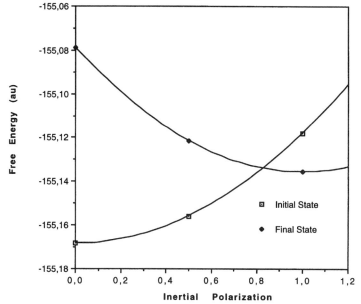

Fig. 17.1. The diabatic nonequilibrium solvent energy surfaces as a function of the inertial polarization.

for the system in which:

(1) The donor is C_2H_4
(2) The acceptor is $C_2H_4^+$
(3) The outer dielectric medium is given by the dielectric properties of acetone ($\epsilon_{op} = 1.841$ and $\epsilon_{st} = 20.7$)

The calculations are performed in such a way that the electronic state of the molecular complex is fixed (either as the initial or the final state) while changing the inertial polarization vector. The inertial polarization vector is changed from that corresponding to the actual electronic state of the molecular complex to the one corresponding to the other electronic state of the molecular complex. The lowest energy of the initial state corresponds to having the inertial polarization vector that corresponds to the actual electronic state of the molecular complex. As soon as the inertial polarization vector is changed, the energy of the initial state increases. Figure 17.1 shows how the energies of the initial and final states change with changes in the inertial polarization vector; for the initial state it changes according to

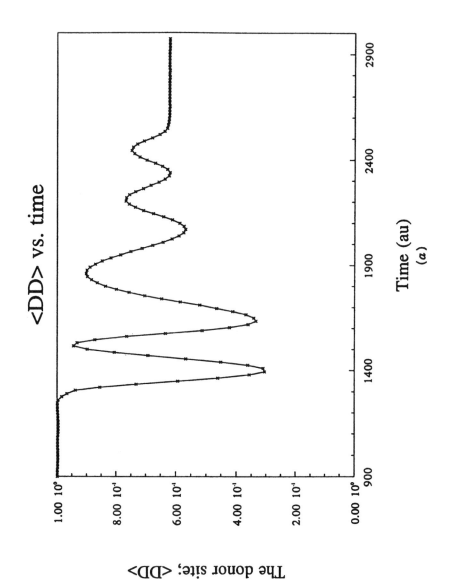

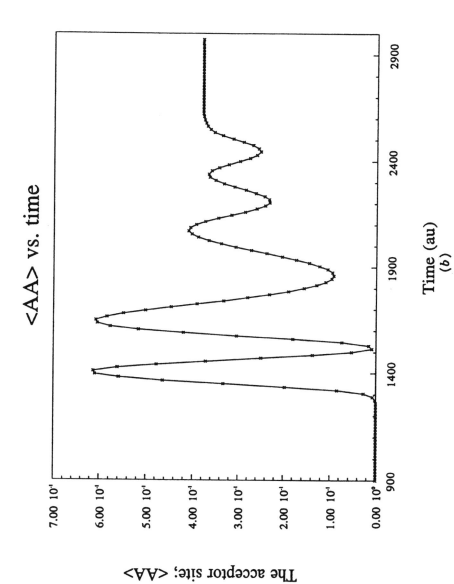

Fig. 17.2. The evolution of the donor (**a**) and acceptor (**b**) sites as a function of time (a.u.).

221

$$\mathbf{P}_{in}(\mathbf{r}) = \mathbf{P}_{in}(\mathbf{r})^{initial} + m(\mathbf{P}_{in}(\mathbf{r})^{final} - \mathbf{P}_{in}(\mathbf{r})^{initial}) \qquad (17.80)$$

where m is varied arbitrarily. The polarization vector $\mathbf{P}_{in}(\mathbf{r})^{initial}$ ($\mathbf{P}_{in}(\mathbf{r})^{final}$) corresponds to the initial (final) state of the electron transfer system. We observe how the diabatic states change with respect to the inertial polarization of the outer medium, and furthermore we are able to locate the transition state, it being the point where the two diabatic curves cross. The latter also enables us to obtain the activation energy.

In Fig. 17.2 we present for the same electron transfer system the change in the donor and acceptor populations during the evolution of the system. In this system

- $\langle a_D^{\dagger} a_D \rangle$ gives the expectation value of finding the electron on the donor
- $\langle a_A^{\dagger} a_A \rangle$ gives the expectation value of finding the electron on the acceptor

The inertial polarization vector is the reaction coordinate and it drives the system from the initial state through the transition state to the final state. We notice that the changes in the expectation values occur over a short time period.

18

ELECTRON TRANSFER COUPLING ELEMENTS

18.1 INTRODUCTION

In order to study the quantum mechanical evolution of electron transfer events, it is necessary to have a method for obtaining the energies of the states involved, along with the coupling elements. The coupling elements describe the electronic interactions between the initial and final states of the electron transfer process. From the numerous excellent reviews in recent years [223–229] one observes that theoretical elucidation of electron transfer events has been of interest and of use to a large scientific community. The many different procedures for calculating the coupling between the initial state (with the electron on the donor) and the final state (with the electron on the acceptor) may be categorized as follows:

(1) Simple one-particle procedures (either an electron or a "hole") [230–243]
(2) Pseudo-two-particle (both electron and hole, which may interfere constructively or destructively) [244–246]
(3) Many-particle methods [247–265]

In order to define the electronic states one can apply electronic structure methods along the lines of:

(1) Extended Hückel methods
(2) Various NDO (neglect of differential overlap) techniques
(3) *Ab initio* techniques

In general the approaches can be divided into three types:

(1) Change in calculated total energies at the transition state structure for the reaction
(2) Application of the frozen orbital method (Koopman's theorem) to estimate the coupling
(3) Direct calculations

One common prerequisite for all such calculations is to perform a careful selection of the basis sets. Cave et al. [252] and Newton [250] have shown the necessity of augmenting the Gaussian-type orbital basis sets of Pople et al. [266–271] and Huzinaga et al. [272, 273] when describing long-range interactions involved in descriptions of electron transfer.

18.2 THEORETICAL METHODS

We consider primarily the theoretical background for performing direct calculations of the electron transfer coupling element. From a given inter- and intramolecular configuration of the donor and acceptor we perform electronic structure calculations on the donor and acceptor complex. As an initial approximation we will assume that it is possible to obtain electronic wave functions where the electron that is going to be transferred is localized on the donor (this being the wave function of the initial state) or on the acceptor (this being the wave function of the final state). These solutions are in general termed broken symmetry solutions [274, 275].

One method for obtaining broken symmetry solutions is to mix the highest occupied and lowest unoccupied molecular orbitals (HOMO and LUMO) and thereby yield an unsymmetric distribution of charge density. Hopefully the minimization of the energy results in broken symmetry solutions leading to initial, bridge, or final states for the electron transfer event. For simplicity we assume that we have obtained two broken symmetry solutions, corresponding to initial and final states (ψ_i and ψ_f) of an electron transfer event. The solutions are obtained at the avoided crossing and have orbitals that are not orthogonal to one another. The two electronic wave functions are not solutions to the same Hamiltonian. In order to proceed we must use a method that can handle the problem of nonorthogonality between two electronic wave functions. One such method is the corresponding orbitals method, which is based upon a pairing theorem of Löwdin [276–278]. The two Hartree-Fock (HF) wave functions ψ_i and ψ_f are given as nonorthogonal Slater determinants of spin orbitals

$$|\psi_i\rangle = \prod_{j=1}^{N_\alpha} a^\dagger_{\alpha, K^i_j} \prod_{k=1}^{N_\beta} a^\dagger_{\beta, K^i_k} |\text{vac}\rangle \tag{18.1}$$

and

$$|\psi_f\rangle = \prod_{j=1}^{N_\alpha} a^\dagger_{\alpha,K^f_j} \prod_{k=1}^{N_\beta} a^\dagger_{\beta,K^f_k} |\text{vac}\rangle \qquad (18.2)$$

where N_x is the number of electrons with spin x, and the operator $a^\dagger_x$ is a creation operator of an electron with spin x. The notations K^i_j and (K^f_j) refer to the jth molecular orbital (MO) of the initial (final) state $i(f)$, and $|\text{vac}\rangle$ refers to the vacuum state. The creation operators are written in ascending order. The molecular orbitals are formed using the chosen basis set of atomic orbitals (AOs). The MOs for the initial and final states are given by

$$|K^i_{x,j}\rangle = a^\dagger_{x,K^i_j} |\text{vac}\rangle \qquad (18.3)$$

$$|K^f_{x,j}\rangle = a^\dagger_{x,K^f_j} |\text{vac}\rangle \qquad (18.4)$$

in which $a^\dagger_{x,k}$ creates an electron with spin x in the kth orbital. The molecular orbitals within the spin orbital sets for the given electronic state are orthogonal. We express the nonorthogonality between the MOs of the different states by the overlap matrix $\mathbf{D}$, whose elements are as follows

$$\mathbf{D}^x_{jk} = \langle K^f_{x,j} | K^i_{x,k} \rangle \neq \delta_{jk} \qquad (18.5)$$

where we have focussed on the spin state x. In the subsequent part of the chapter we will not specify the spin state; however, it is understood that α and β MOs must be dealt with separately. The next step is to perform unitary transformations on the sets of spin orbitals $|K^i\rangle$ and $|K^f\rangle$

$$|\hat{K}^i_j\rangle = \mathbf{V}|K^i_j\rangle \qquad (18.6)$$

$$|\hat{K}^f_j\rangle = \mathbf{U}|K^f_j\rangle \qquad (18.7)$$

which leads to a transformation of the overlap matrix $\mathbf{D}$

$$\langle \hat{K}^f | \hat{K}^i \rangle = \hat{\mathbf{d}} = \mathbf{U}^\dagger \mathbf{D} \mathbf{V} \qquad (18.8)$$

We obtain the two transformation matrices ($\mathbf{U}$ and $\mathbf{V}$) and the (diagonal) eigenvalue matrix $\mathbf{\Lambda}$ from the equations

$$\mathbf{D}^\dagger \mathbf{D} \mathbf{V} = \mathbf{V} \mathbf{\Lambda} \qquad (18.9)$$

and

$$\mathbf{U} = \mathbf{DV}\mathbf{\Lambda}^{-1/2} \tag{18.10}$$

From Eq. (18.9) we have

$$\mathbf{\Lambda} = \mathbf{V}^\dagger \mathbf{D}^\dagger \mathbf{DV} \tag{18.11}$$

and proceeding from Eq. (18.8) we have

$$\hat{\mathbf{d}} = \mathbf{U}^\dagger \mathbf{DV} = \mathbf{\Lambda}^{-1/2}\mathbf{V}^\dagger \mathbf{D}^\dagger \mathbf{DV} = \mathbf{\Lambda}^{1/2} \tag{18.12}$$

which has the required diagonal form. The transformations have transformed the original overlap matrix $\mathbf{D}$ into a diagonal overlap matrix containing elements with values in the interval [0, 1]. The overlap integral between the two electronic wave functions for the electron transfer event (ψ_i and ψ_f) is then given by

$$S_{fi} = (\det\mathbf{U})(\det\mathbf{V}^\dagger) \prod_{j=1}^N \hat{d}_{jj} \tag{18.13}$$

Next we consider how the Slater-Condon rules will be modified. We begin with an energy operator

$$\Omega_{op} = \Omega^{(0)} + \Omega^{(1)} + \Omega^{(2)} \tag{18.14}$$

Here, $\Omega^{(0)}$ is the nuclear repulsion operator, $\Omega^{(1)}$ is the one-electron operator, and $\Omega^{(2)}$ is the two-electron operator. Utilizing a strategy similar to that for the overlap S_{fi} we can obtain the various components of the energy as follows:

$$\Omega_{if}^{(0)} = \Omega^{(0)}(\det\mathbf{U})(\det\mathbf{V}^\dagger) \prod_{j=1}^N \hat{d}_{ij} \tag{18.15}$$

$$\Omega_{if}^{(1)} = (\det\mathbf{U})(\det\mathbf{V}^\dagger) \sum_{k=1}^N \langle \hat{K}_k^f | \omega | \hat{K}_k^i \rangle \prod_{j \neq k}^N \hat{d}_{jj} \tag{18.16}$$

$$\Omega_{if}^{(2)} = (\det\mathbf{U})(\det\mathbf{V}^\dagger) \sum_{j<k}^N \langle \hat{K}_j^f \hat{K}_k^f | \omega(1,2)(1 - P_{12}) | \hat{K}_j^i \hat{K}_k^i \rangle \prod_{m \neq j,k}^N \hat{d}_{mm}$$

$$\tag{18.17}$$

in which ω is the kinetic plus nuclear attraction energy operator, $\omega(1,2)$ is $1/R_{12}$ (the distance between the electrons in atomic units), and P_{12} is the exchange operator

$$P_{12}|\hat{K}_j^i \hat{K}_k^i\rangle = |\hat{K}_k^i \hat{K}_j^i\rangle \tag{18.18}$$

Next we introduce the following elements:

(1) The energy of the initial state $H_{ii} = \langle \psi_i |\Omega_{\text{op}}| \psi_i \rangle$

(2) The energy of the final state $H_{ff} = \langle \psi_f |\Omega_{\text{op}}| \psi_f \rangle$

(3) The coupling between the initial and final state $H_{fi} = \langle \psi_f |\Omega_{\text{op}}| \psi_i \rangle$

which enable us to obtain the coupling element T_{fi} (or one half the splitting between the symmetric and antisymmetric solutions at the transition state) as

$$T_{fi} = \frac{H_{fi} - S_{fi}H_{ii}}{1 - S_{fi}^2} \tag{18.19}$$

For the two-state CI (configuration interaction) problem with nonorthogonal wave functions, one has an off-diagonal element that is the T_{fi} element. For direct calculation of coupling elements between initial and final states for the ET reaction, another procedure has been presented that employs the Löwdin symmetric orthogonalization scheme for taking account of the non-orthogonality [259–264]. With this procedure a Löwdin symmetric orthogonalization between the initial and final wavefunctions is performed and one calculates the coupling elements by a similar method to what has already been described.

The most frequently used methods for obtaining the electron transfer coupling elements are indirect methods. One of these, the ΔSCF method, involves the calculation of total energies at a given geometry of the ground and first excited states involved in the electron transfer reaction. Assuming that the two states involved in the electron transfer reaction are well separated from all other states, one finds that the energy difference between the two states at the transition state configuration is twice the coupling element for the electron transfer reaction. An estimation of this energy difference can be achieved through the use of Koopman's theorem [279], which makes the assumption that the spin orbitals in the $(N \pm 1)$-electron states are identical with those of the N-electron state. Along these lines we can approximate the energy splitting between the two states as the difference in energy between the HOMO and the LUMO, or the initial MO and the final MO.

Let us consider the scheme for indirect calculation of the coupling element by utilizing the diabatic representation for the electronic wave functions (ψ_i and ψ_f). They represent electronic states where the excess electron is either located on the donor (ψ_i) or acceptor (ψ_f). Within the adiabatic representation, the electronic wave functions are given by the ansatz

$$\Psi(Q) = C_i(Q)\psi_i(Q) + C_f(Q)\psi_f(Q) \tag{18.20}$$

where Q denotes a generalized reaction coordinate. We obtain the energies for the two states in the adiabatic representation according to

$$H(Q)\Psi(Q) = \Psi(Q)E(Q) \tag{18.21}$$

And by making use of Eq. (18.20) we get the following secular equation

$$\begin{pmatrix} H_{ii} - E & H_{if} - S_{if}E \\ H_{fi} - S_{fi}E & H_{ff} - E \end{pmatrix} \begin{pmatrix} C_i \\ C_f \end{pmatrix} = \begin{pmatrix} 0 \\ 0 \end{pmatrix} \tag{18.22}$$

where we have defined the following quantities:

(1) The overlap $S_{fi}(Q)$, which depends on the reaction coordinate and is determined as $S_{fi}(Q) = S_{if}(Q) = S_{if} = \langle \psi_i(Q)|\psi_f(Q)\rangle$
(2) The expectation value of the Hamiltonian with respect to state j, which is given as $H_{jj}(Q) = H_{jj} = \langle \psi_j(Q)|H(Q)|\psi_j(Q)\rangle$
(3) The coupling between the two states ψ_i and ψ_f, this element being obtained as $H_{fi}(Q) = H_{if}(Q) = H_{if} = \langle \psi_f(Q)|H(Q)|\psi_i(Q)\rangle$

We are then able to express the energies in terms of these quantities and obtain

$$E(Q)^{\pm} = \frac{H_{ii} + H_{ff} - 2H_{if}S_{if}}{2(1 - S_{if}^2)} \pm \frac{\sqrt{D}}{2(1 - S_{if}^2)} \tag{18.23}$$

where

$$D = (H_{ii} - H_{ff})^2 + 4[H_{if}^2 - H_{if}S_{if}(H_{ii} + H_{ff}) + S_{if}^2 H_{ii}H_{ff}] \tag{18.24}$$

Within this model we are able to get, for a given nuclear configuration, the energy difference between the two adiabatic states, and we find that

$$\Delta E(Q) = \frac{\sqrt{D}}{2(1 - S_{if}^2)} \tag{18.25}$$

We assume that we are able to locate the transition state configuration and denote the reaction coordinate associated with this configuration Q^*. At Q^* the diabatic energy surfaces cross and we have

$$H_{ii}(Q^*) = H_{ff}(Q^*) \qquad (18.26)$$

which leads to the expression

$$D = 4(H_{if}(Q^*) - S_{if}(Q^*)H_{ii}(Q^*))^2 \qquad (18.27)$$

We therefore determine the energy splitting to be

$$\Delta E(Q) = 2\,\frac{H_{if} - S_{if}H_{ii}}{(1 - S_{if}^2)} \qquad (18.28)$$

with the requirement that all the elements H_{if}, S_{if}, and H_{ii} are evaluated at the transition state configuration denoted Q^*.

An important point to note [from Eqs. (18.24–18.28)] is that the energy splitting is equal to twice the coupling element for the ET reaction at only those nuclear configurations for which $H_{ii}(Q) = H_{ff}(Q)$, i.e., the transition state configuration(s). Therefore, for the indirect calculations of the electron transfer coupling elements, it is extremely important to use the appropriate nuclear configuration; otherwise, it is necessary to employ the full expression for D found in Eq. (18.24). The transition state is given by the topology of the energy surfaces and the determination of the transition state configurations is not a trivial matter. In a lot of ways, indirect methods remain highly inaccurate and it is not always clear which nuclear configuration one should use for obtaining the coupling element. The direct methods require some nonstandard techniques and implementation of computer code, but the method is of a much more rigorous character.

18.2.1 Numerical Examples

In Figure 18.1 we present Hartree-Fock calculations of the electron transfer coupling element between $C_2H_4^+$ as the acceptor and C_2H_4 as the donor. The electron transfer coupling element has been calculated using the direct method from Eq. (18.19) as a function of distance, and for various bridge molecules placed between the donor and acceptor. We observe that:

(1) The electron transfer coupling element decreases exponentially with the donor-acceptor distance.

(2) At a given donor–acceptor distance, the coupling element increases in the presence of a molecule between the donor and acceptor.

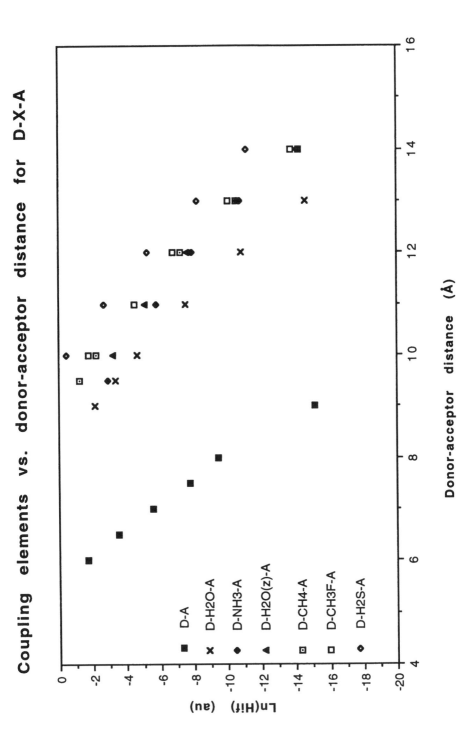

Fig. 18.1. Electron transfer coupling elements as function of donor—acceptor distance. Various bridge molecules are included.

The latter point clearly demonstrates the possibility of designing efficient or inefficient bridging groups between donor and acceptor sites.

18.3 METHODS FOR ELECTRON COUPLING ELEMENTS IN PROTEINS

Within the last 15 years, electron transfer in or between proteins has become a very active research area. Attempts to calculate electron transfer coupling elements have involved:

(1) Calculation of the number of bonds or paths between the redox centers [280]

(2) Application of superexchange methods [281–283]

(3) Indirect methods, as described in the previous section [284, 285]

(4) Direct methods, as described in the previous section [286]

In this section we describe a superexchange model, introduced in ref. [287], in which the model system consists of:

(1) A donor site with S_D spin orbitals and N_D electrons

(2) An assembly of bridge sites, which are numbered $J = 1, M$, and for which bridge site i has S_{B_i} spin orbitals and N_{B_i} electrons

(3) An acceptor site having S_A spin orbitals and N_A electrons

We assume that it is possible to obtain the MO basis by symmetrical orthogonalization [276] from all the orbitals on the donor, acceptor, and bridge sites. Furthermore, we let the electron number operators for the donor, acceptor and bridge groups be given by

$$\mathbf{N}_D = \sum_{s \in D} a_s^\dagger a_s \tag{18.29}$$

$$\mathbf{N}_A = \sum_{s \in A} a_s^\dagger a_s \tag{18.30}$$

$$\mathbf{N}_{B_i} = \sum_{s \in B_i} a_s^\dagger a \qquad \forall_i = 1, \dots, M \tag{18.31}$$

where the $a_s^\dagger$ (a_s) is the electron creation (annihilation) operator. With this definition the total number operator can be written as

$$\mathbf{N}_T = \mathbf{N}_D + \mathbf{N}_A + \sum_{i=1}^{M} \mathbf{N}_{B_i} \tag{18.32}$$

The initial, final and all the bridge states are assumed to be eigenstates of the number operators and we represent these states as

- The initial state:

$$\Phi_D = |1,\ldots,N_D,N_D+1; 1,\ldots,N_{B_1}; 1,\ldots,N_{B_2};\ldots; 1,\ldots,N_{B_N}; 1,\ldots,N_A\rangle$$

- The final state:

$$\Phi_A = |1,\ldots,N_D; 1,\ldots,N_{B_1}; 1,\ldots,N_{B_2};\ldots; 1,\ldots,N_{B_N}; 1,\ldots,N_A,N_A+1\rangle$$

- The intermediate state corresponding to electron localization on the pth bridge site:

$$\Phi_{B_p} = |1,\ldots,N_D; 1,\ldots,N_{B_1};\ldots; 1,\ldots,N_{B_p},N_{B_p}+1;\ldots; 1,\ldots,N_{B_N}; 1,\ldots,N_A\rangle$$

- The intermediate state corresponding to hole transport involving the pth bridge site:

$$\Phi_{B_p^h} = |1,\ldots,N_D,N_D+1; 1,\ldots,N_{B_1};\ldots; 1,\ldots,N_{B_p}-1;\ldots;$$
$$1,\ldots,N_{B_N}; 1,\ldots,N_A,N_A+1\rangle$$

The total Hamiltonian is divided into a zero-order Hamiltonian, H^o, and a perturbation V. The perturbation couples the different sites and for the zero-order Hamiltonian we have that

$$H^o\Phi_D = E_D\Phi_D \tag{18.33}$$

$$H^o\Phi_{B_i} = E_{B_i}\Phi_{B_i} \tag{18.34}$$

$$H^o\Phi_A = E_A\Phi_A \tag{18.35}$$

where E_D, E_{B_i}, and E_A are the corresponding eigenvalues. The next step will be to solve the time-dependent Schrödinger equation

$$i\hbar \, \frac{\partial \Psi(t)}{\partial t} = (H^o + V)\Phi(t) \tag{(18.36)}$$

with the initial condition that

$$\Psi(t = -\infty) = \Phi_D \tag{18.37}$$

The initial condition states that the electron to be transferred is initially located on the donor site. We would like to know the probability of having the transferable electron on the acceptor site at $t = 0$. Making use of the interaction representation allows us to rewrite the operators and wave functions as

$$\Psi_I(t) = \exp\left(\frac{iH^ot}{\hbar}\right)\Psi(t) \tag{18.38}$$

$$A_I(t) = \exp\left(\frac{iH^ot}{\hbar}\right)A\exp\left(-\frac{iH^ot}{\hbar}\right) \tag{18.39}$$

where we have used a generic operator A to illustrate the representation. Within the interaction representation, the time-dependent Schrödinger equation becomes

$$i\hbar\,\frac{\partial\Psi_I(t)}{\partial t} = V_I(t)\Psi_I(t) \tag{18.40}$$

and formally the solution at $t = 0$ can be written as

$$\Psi_I(t = 0) = \Phi_D - \frac{i}{\hbar}\lim_{\delta \to 0^+}\int_{-\infty}^{0} dt'\exp\left(-\frac{\delta|t'|}{\hbar}\right)V_I(t')\Psi_I(t') \tag{18.41}$$

Iteratively we obtain the following expression for the resulting wave function, Ψ_{res}

$$\begin{aligned}\Psi_{res} = \Psi_I(t = 0) = \Phi_D &+ \lim_{\delta \to 0^+}\frac{1}{E_D - H^o + i\delta}V\Psi_{res}\\[1em] = \Phi_D &+ \lim_{\delta \to 0^+}\frac{1}{E_D - H^o + i\delta}V\Phi_D\\[1em] &+ \lim_{\delta \to 0^+}\frac{1}{E_D - H^o + i\delta}V\frac{1}{E_D - H^o + i\delta}V\Phi_D + \cdots\end{aligned} \tag{18.42}$$

We can thus use the iterative solution to obtain the transition matrix element (the coupling element or the scattering amplitude) between the donor and acceptor sites

$$T_{DA} = \langle \Phi_A | V | \Psi_{res} \rangle = \langle \Phi_A | V | \Psi_D \rangle + \lim_{\delta \to 0^+} \sum_{B_i, \alpha} \frac{\langle \Phi_A | V | \Phi_{B_i, \alpha} \rangle \langle \Phi_{B_i, \alpha} | V | \Phi_D \rangle}{(E_D - E_{B_i, \alpha} + i\delta)}$$

$$+ \cdots + \lim_{\delta \to 0^+} \sum_{B_i, \alpha} \sum_{B_j, \beta} \cdots \sum_{B_t, \tau}$$

$$\cdot \frac{\langle \Phi_A | V | \Phi_{B_i, \alpha} \rangle \langle \Phi_{B_i, \alpha} | V | \Phi_{B_j, \beta} \rangle \cdots \langle \Phi_{B_t, \tau} | V | \Phi_D \rangle}{(E_D - E_{B_i, \alpha} + i\delta)(E_D - E_{B_j, \beta} + i\delta) \dots (E_D - E_{B_t, \tau} + i\delta)} \quad (18.43)$$

where the first term gives the direct coupling between the donor and acceptor sites. The first summation accounts for the coupling between the donor and acceptor sites via any one of the M bridges that the electron can pass through. The last summation allows the donor and acceptor sites to interact via all the M bridges. The Greek letters denote the different states on the bridges.

In a situation where it is only nearest-neighbor interactions that are of importance, we are left with the term that represents the number of bridges between the two sites, (let that be M), and we have

$$T_{DA} = \sum_{B_1, \alpha} \sum_{B_2, \beta} \cdots \sum_{B_M, \tau} \frac{\langle \Phi_D | V | \Phi_{B_1, \alpha} \rangle \langle \Phi_{B_1, \alpha} | V | \Phi_{B_2, \beta} \rangle \dots \langle \Phi_{B_M, \tau} | V | \Phi_A \rangle}{(E_D - E_{B_1, \alpha})(E_D - E_{B_2, \beta}) \dots (E_D - E_{B_M, \tau})}$$

$$(18.44)$$

and we can reduce this further when only a single electronic state on each bridge is of importance

$$T_{DA} = \frac{\langle \Phi_D | V | \Phi_{B_1} \rangle \langle \Phi_{B_1} | V | \Phi_{B_2} \rangle \dots \langle \Phi_{B_M} | V | \Phi_A \rangle}{(E_D - E_{B_1})(E_D - E_{B_2}) \dots (E_D - E_{B_M})} \quad (18.45)$$

These expressions, or modifications thereof, have been used to establish guidelines for electron transport in proteins [287–290].

18.3.1   Numerical Examples

Here we present electron transfer elements for different electron transfer pathways in the photosynthetic blue single-copper protein plastocyanin (PC), see refs. [287, 290] for further details. Application of Eq. (18.45) along with extended Hückel calculations for five different electron transfer pathways show rather different properties for electron transport (see Table 18.1). Pathway 5 is the direct pathway from the redox center to the outside redox partner. The four

TABLE 18.1. The Electron Transfer Coupling Elements for Five Different Electron Transfer Pathways in Plastocyanin Plastocyanin

Pathway	T_{DA} (cm^{-1})
1	1×10^{-9}
2	3×10^{-4}
3	6×10^{-2}
4	6×10^{-4}
5	0.4

other pathways are through the protein and the distances between the redox centers in these cases are larger by about 5 to 10 Å than the distance between the redox centers for pathway 5. The most efficient pathway (pathway 3) through the protein involves the two amino acids Cys-83 and Tyr-83. It is clear that the protein environment is rather selective with respect to transport of electrons.

19

ELECTRON TRANSFER REACTIONS COUPLED TO A QUANTUM MECHANICAL RADIATION FIELD

The topic of this chapter is the interaction between molecules and quantized fields, in particular, strong fields. The chemical reaction we will consider is a charge transfer reaction induced by a laser field. In the case of strong-light fields, investigations on atoms have taken place both experimentally and theoretically [291–295]. The problem of molecules in strong fields has been addressed by few researchers [296–307], and a systematic theoretical treatment does not appear to exist as yet. The focus of these studies has been radiation in the infrared region [298, 301, 305, 308–311]. The molecular problem contains a greater number of degrees of freedom than the atomic one, and it is reasonable to expect that the resulting dynamics from the laser–molecule interaction would be quite complex. It is also expected that the results from models describing the molecular system plus radiation field will depend on several parameters related to the properties of the laser pulse:

(1) The intensity of the laser field
(2) Whether the laser is pulsed or continuous wave
(3) The shape of the laser pulse
(4) The duration of the laser pulse

However, it is not straightforward to characterize experimentally the laser pulses in the case of laser fields of high intensities, and we refer the reader to ref. [297] for further details.

For a general introduction to quantum optics we refer the reader to refs. [294, 312] where investigations concerning atomic systems interacting with a photon field have been reviewed in great detail. Among the effects that have been postulated and demonstrated are:

(1) Laser cooling of atoms
(2) Resonance fluorescence
(3) Photon correlations in intense resonant or quasi-resonant laser beams

In this chapter we illustrate a model for describing the effects of a quantum mechanical radiation field on a two-level molecular system and furthermore present a method for calculating the quantities needed for investigating the evolution of the molecular system coupled to the radiation field. Introducing a quantum mechanical radiation field into the models makes it possible to describe effects such as:

(1) Vacuum fluctuations
(2) Spontaneous emission
(3) Quantum beats
(4) Laser linewidths

We consider here a solvated molecular system interacting with a laser field. The molecular system is surrounded by a dielectric medium. The dielectric medium is described by two polarization vectors, the inertial polarization vector and the optical polarization vector. The sum of these two vectors gives the total polarization vector. The optical polarization vector is related to the optical dielectric constant, whereas the total polarization vector is related to the static dielectric constant. The solvated molecular system will undergo laser-field-induced intramolecular charge transfer, and we let the monochromatic laser field be resonant or quasi-resonant with the charge transfer transition. The molecular system interacts with the outer solvent, along with the laser field and the reservoir of initially empty modes of the quantum mechanical radiation field. We term the radiation mode initially occupied with photons of a specific energy and polarization the laser mode, and we assume that the electromagnetic radiation field is confined in an electromagnetic cavity having a volume V. The number of photons in the quantum mechanical radiation field corresponds to the mesoscopic region bridging the gap between a microscopic and a macroscopic system. In general, one define a microscopic system to be one where a molecule interacts with a few photons contained within a microcavity. On the other hand a macroscopic system would contain well over 100 photons. A discussion of the differences between the three systems can be found in ref. [293].

Actual calculations of the necessary energies and coupling elements needed for solving the evolution equations of the combined system consisting of solvated para-nitroaniline (PNA) and electromagnetic fields can be performed with the response method described in Chapter 15 on molecular properties of solvated molecules. The response method and associated calculations enable us to obtain these quantities when the molecule is solvated.

19.1   THE MODEL

Once again we apply Ehrenfest's equation to describe the evolution of the total system, i.e., the molecular system, the radiation field and the outer solvent. Thus,

$$i \frac{d}{dt} \langle \Omega \rangle = \langle [\Omega, H] \rangle \tag{19.1}$$

The expectation value of the operator $\langle \Omega \rangle$ is given by

$$\langle \Omega \rangle = \text{Tr}[\Omega \rho] \tag{19.2}$$

and the expectation value of the commutator between the operator and the Hamiltonian is

$$\langle [\Omega, H] \rangle = \text{Tr}\{[\Omega, H]\rho\} \tag{19.3}$$

where ρ is the density operator of the total system.

The changes in the expectation values of appropriate operators will tell us how the total system is evolving. Appropriate operators in our case will be:

(1) The operators $a_r^{\dagger} a_s$, $a_s^{\dagger} a_r$. These operators operate on the electronic degrees of freedom and the operator $a_r^{\dagger}$ (a_r) creates (destroys) an electron in level r. The combined effect of two of these operators, for example $a_r^{\dagger} a_s$, is to move the electron from level s to level r.

(2) The operators $b_k^{\dagger}$, b_k. These operators operate on the degrees of freedom of the outer medium (the solvent). The effect of these operators is that $b_k^{\dagger}$ (b_k) creates (destroys) a phonon in the kth mode in the solvent. This set of operators is divided into two groups:

   (i) Those operators corresponding to the modes in the dielectric medium responsible for the optical polarization vector (the high-frequency modes, these modes being indicated by the superscript *high*)

   (ii) Operators corresponding to all the other modes; the low frequency modes, (indicated by the superscript *low*), that are responsible for the inertial polarization.

(3) The operators $B_n^{\dagger}$, B_n. The operators $B_n^{\dagger}$, B_n operate on the intramolecular degrees of freedom of the molecule. The effect of the operator $B_n^{\dagger}$ (B_n) is that it creates (destroys) a vibration in the nth vibrational mode of the molecule.

(4) The operators $d_{p\lambda}^{\dagger}$, $d_{p\lambda}$. These operators act on the photon degrees of freedom, the radiation field that includes the laser field and the radiation reservoir. The operator $d_{p\lambda}^{\dagger}$ ($d_{p\lambda}$) creates (destroys) a photon in the radiation field with momentum vector $\mathbf{p}$ and polarization λ.

For this manifold of operators we formally obtain the following set of equations of motion:

$$i \frac{d}{dt} \langle a_r^\dagger a_s \rangle = \langle [a_r^\dagger a_s, H] \rangle \tag{19.4}$$

$$i \frac{d}{dt} \langle b_k^{\text{high}} \rangle = \langle [b_k^{\text{high}}, H] \rangle \tag{19.5}$$

$$i \frac{d}{dt} \langle b_k^{\text{low}} \rangle = \langle [b_k^{\text{low}}, H] \rangle \tag{19.6}$$

$$i \frac{d}{dt} \langle B_n \rangle = \langle [B_n, H] \rangle \tag{19.7}$$

$$i \frac{d}{dt} \langle d_{p\lambda} \rangle = \langle [d_{p\lambda}, H] \rangle \tag{19.8}$$

Next we must consider the Hamiltonian for the system.

19.2   THE HAMILTONIAN

The total system consists of a molecule immersed in a condensed phase, a solution, interacting with a radiation field. The solution is represented as a phonon bath, and the radiation field is given by the vacuum photon field and an applied laser field. The inclusion of the vacuum field enables us to describe vacuum fluctuations. A physical phenomenon related to the vacuum field is the spontaneous emission of photons by excited molecular states. The molecular degrees of freedom are represented by the electronic and vibrational degrees of freedom of the molecule.

The Hamiltonian for the total system is given as

$$H = H_e + H_v + W_{e-v} + H_s + W_{o-s} + H_r + W_{o-r} \tag{19.9}$$

where the different terms are

- H_e, the many-electron Hamiltonian for the electronic degrees of freedom of the molecular system
- H_v, the Hamiltonian for the intramolecular degrees of freedom contained in the molecular system
- W_{e-v}, the vibronic coupling operator, which describes the coupling between the electronic and intramolecular degrees of freedom
- H_s, the Hamiltonian for the outer solvent

- W_{o-s}, the interaction operator, which takes into account the electrostatic interactions between the molecular charge distribution and the outer solvent
- H_r, the Hamiltonian for the radiation field
- W_{o-r}, which takes care of the interactions between the radiation field and the molecular system

It is important to note that the radiation field is treated quantum mechanically and no separation of the laser mode and reservoir modes is made.

In second quantization we write the many-electron Hamiltonian as

$$H_e = \sum_{r,s} h_{rs} a_r^\dagger a_s + \frac{1}{2} \sum_{r,s,t,u} (rs|tu) a_r^\dagger a_t^\dagger a_u a_s \qquad (19.10)$$

where we sum over molecular orbitals. The integrals h_{rs} and $(rs|tu)$ denote the usual one-electron and two-electron integrals.

The Hamiltonian for the intramolecular degrees of freedom is given by

$$H_v = \sum_n \Omega_n B_n^\dagger B_n \qquad (19.11)$$

where we have utilized the harmonic approximation for the vibrations. We have that $\{\Omega_n\}$ and $\{B_n^\dagger, B_n\}$ are the frequencies and the boson operators for the modes, respectively. The boson operators obey the rules of ordinary boson algebra.

The Hamiltonian that takes account of the vibronic coupling operator is obtained by performing a first order Taylor expansion of the many-electron Hamiltonian in terms of the intramolecular degrees of freedom around a given intramolecular configuration. Therefore, we write this operator as

$$W_{e-v} = \sum_n \left(\frac{\partial H_e(\{r_i\}, \{Q_n\}, \{k\})}{\partial Q_n} \right)_{Q_n = Q_n^o} (Q_n - Q_n^o)$$

$$= \sum_n A_n H_n^{(1)} (B_n^\dagger + B_n) \qquad (19.12)$$

where

- $\{r_i\}$ represents the electronic degrees of freedom of the molecular compounds
- $\{Q_n\}$ represents the nuclear degrees of freedom of the molecular compounds

- $\{k\}$ represents the bulk solvent modes, both the high- and low-frequency modes

We find that

$$H_n^{(1)} = \left(\frac{\partial H_e(\{r_i\}, \{Q_n\}, \{k\})}{\partial Q_n} \right)_{Q_n = Q_n^o} \tag{19.13}$$

$$(Q_n - Q_n^o) = \frac{B_n^\dagger + B_n}{\sqrt{2M_n\Omega_n}}$$

$$= \Delta Q_n(B_n^\dagger + B_n) \tag{19.14}$$

where ΔQ_n and M_n are the zero point displacement along the mode n and the reduced mass of the mode n, respectively. The coupling factor A_n is given by

$$A_n = \sqrt{2M_n\Omega_n} \tag{19.15}$$

The outer solvent is represented by the frequencies $\{\omega_k\}$ and the boson operators $\{b_k^\dagger, b_k\}$ for the modes. The boson operators satisfy ordinary boson algebra. We have the following Hamiltonian for the outer solvent

$$H_s = \sum_k \omega_k b_k^\dagger b_k \tag{19.16}$$

Interactions between the molecular charge distribution and the solution are given by

$$W_{o-s} = \sum_k \{\alpha_k M_k b_k^\dagger + \alpha_k^* M_k^\dagger b_k\} \tag{19.17}$$

where α_k, α_k^* are parameters describing the dielectric interaction between the molecular charge distribution and the outer medium. The operators $\{M_k^\dagger, M_k\}$ are the charge moment operators of the molecular charge distribution. We obtain those by performing multipole expansion of the molecular charge distribution.

The two components in the Hamiltonian for the photon part are presented in the appendix B on quantum mechanical radiation fields. The following Hamiltonian

$$H_r = \sum_{p\lambda} pc\, d_{p\lambda}^\dagger d_{p\lambda} \tag{19.18}$$

where c is the speed of light and $p = |\mathbf{p}|$ describes the radiation field. The radiation field is the combined laser field and vacuum radiation field.

The operator that describes the interaction between the photons and the molecule is represented by [313]

$$W_{o-r} = -\frac{1}{c} \int A(\mathbf{r}) \cdot J(\mathbf{r}) \, d\mathbf{r} \qquad (19.19)$$

where $A(\mathbf{r})$ is the vector potential of the radiation field in the Coulomb gauge and $J(\mathbf{r})$ is the current density operator of the molecular system. Further simplification can be reached by writing the vector potential in the Linderberg-Öhrn notation [313]

$$A(\mathbf{r}) = \sum_{\mathbf{p}\lambda} \sqrt{\frac{2\pi c}{pV}} \, [d_{\mathbf{p}\lambda}^{\dagger} - (-1)^{\lambda} \, d_{\mathbf{p}\lambda}] \mathbf{n}(\mathbf{p}) e^{i\mathbf{p}\cdot\mathbf{r}} \qquad (19.20)$$

The molecular current density operator is then given by

$$
\begin{aligned}
j_{\lambda}(\mathbf{p}) &= \mathbf{n}(\mathbf{p}) \cdot \int d\mathbf{r} \, \mathbf{J}(\mathbf{r}) e^{-i\mathbf{p}\cdot\mathbf{r}} \\
&= \sum_{uv} (u|j_{\lambda}(\mathbf{p})|v) a_u^{\dagger} a_v
\end{aligned}
\qquad (19.21)
$$

where the summation is over all molecular orbitals, u, v. This allows us to rewrite the molecule–radiation field interaction operator as

$$W_{o-r} = -\sum_{\mathbf{p}\lambda} \sqrt{\frac{2\pi}{cpV}} \, [d_{\mathbf{p}\lambda}^{\dagger} j_{\lambda}(\mathbf{p}) + d_{\mathbf{p}\lambda} j_{\lambda}^{\dagger}(\mathbf{p})] \qquad (19.22)$$

Having established the Hamiltonian for the system, the next step is to find a representation for the state.

19.3   THE TIME-DEPENDENT STATISTICAL DENSITY OPERATOR

The time-dependent statistical density operator is given by the direct product

$$\rho_{\text{tot}}(t) = \rho_m(t) \times \rho_s(t) \times \rho_r(t) \qquad (19.23)$$

where

- $\rho_m(t)$ is the statistical density operator for the molecular system
- $\rho_s(t)$ is the statistical density operator for the outer solvent
- $\rho_r(t)$ is the statistical density operator for the radiation field

The statistical density operator for the outer solvent is given by

$$\rho_s(t) = e^{-iS(t)}\rho_s(o)e^{iS(t)} \tag{19.24}$$

where $\rho_s(o)$ is the statistical density operator for the outer medium when uncoupled from the molecular charge distribution, and is assumed to be a Boltzmann distribution at temperature T

$$\rho_s(o) = \frac{\exp(-\beta \sum_k \omega_k b_k^\dagger b_k)}{\text{Tr}[\exp(-\beta \sum_k \omega_k b_k^\dagger b_k)]} \tag{19.25}$$

A unitary transformation takes account of the changes in the statistical density operator for the outer solvent due to interactions between the molecular charge distribution and the outer medium. The transformation operator S is given as

$$S(t) = \sum_k (X_k(t)b_k^\dagger + X_k^*(t)b_k) \tag{19.26}$$

The values of the time-dependent parameters $X_k(t)$, $X_k^*(t)$ are determined by the strength of the interactions between the molecular charge distribution and the outer solvent.

We let the radiation field be represented by the following density operator

$$\rho_r(t) = |r(t)\rangle\langle r(t)| \tag{19.27}$$

where

$$|r(t)\rangle = e^{iZ(t)} \prod_{p\lambda} \frac{(d_{p\lambda}^\dagger)^{n_{p\lambda}}}{\sqrt{(n_{p\lambda})!}} |vac\rangle \tag{19.28}$$

where the unitary transformation is given as

$$Z(t) = \sum_{p\lambda} (\xi_{p\lambda}(t)d_{p\lambda}^\dagger + \xi_{p\lambda}^*(t)d_{p\lambda}) \tag{19.29}$$

and where $|vac\rangle$ is the vacuum state for the electromagnetic field and $n_{p\lambda}$ is the occupation number for a state of the electromagnetic field with wave vector **p** and polarization λ. The purpose of the unitary transformation is to take care of the dynamical interaction between the molecular system and the electromagnetic field. Let us consider the situation where we only consider the interactions between the photons and an electronic state $|\Theta\rangle$; we are then able to write the total energy as a function of the parameters $\xi_{p\lambda}$ and $\xi'_{p\lambda}$ [221, 314, 315]

$$E(\xi_{p\lambda}) = \langle\Theta|H_e|\Theta\rangle - \sum_{p\lambda} pc\, \xi_{p\lambda}(t)\xi'_{p\lambda}(t)$$

$$-i \sum_{p\lambda} \sqrt{\frac{2\pi}{pcV}} \{\xi_{p\lambda}(t)\langle\Theta|j_\lambda^\dagger(\mathbf{p})|\Theta\rangle$$

$$-\xi'_{p\lambda}(t)\langle\Theta|j_\lambda(\mathbf{p})|\Theta\rangle\} \tag{19.30}$$

If we require that the energy functional be stationary with respect to variations in the parameters $\xi_{p\lambda}(t)$ and $\xi'_{p\lambda}(t)$ we obtain

$$E(\xi_{p\lambda}) = \langle\Theta|H_e|\Theta\rangle$$

$$- \sum_{p\lambda} \left(\frac{2\pi}{p^2c^2V}\right) \langle\Theta|j_\lambda^\dagger(\mathbf{p})|\Theta\rangle\langle\Theta|j_\lambda(\mathbf{p})|\Theta\rangle \tag{19.31}$$

where the second term shows the level shift due to the emission and absorption of virtual photons.

For the molecular system, the statistical density operator is given by

$$\rho_m(t) = |\Phi\rangle\langle\Phi| \times \rho_v(t) \tag{19.32}$$

where

- $|\Phi\rangle$ is the electronic wave function for the molecular system inside the solvent cavity
- $\rho_v(t)$ is the statistical density operator for the intramolecular degrees of freedom for the molecular compounds inside the solvent cavity

The statistical density operator for the intramolecular degrees of freedom of the molecular compounds inside the solvent cavity is given as

$$\rho_v(t) = e^{-iV(t)}\rho_v(o)e^{iV(t)} \tag{19.33}$$

where $\rho_v(o)$ is the statistical density operator for a Boltzmann distribution at temperature T describing the uncoupled intramolecular vibrations

$$\rho_v(o) = \frac{\exp(-\beta \sum_n \omega_n B_n^\dagger B_n)}{\text{Tr}[\exp(-\beta \sum_n \omega_n B_n^\dagger B_n)]} \tag{19.34}$$

and the unitary transformations performed by the operator $V(t)$ change the statistical density operator for the Boltzmann distribution, $\rho_v(o)$, into the statistical density operator $\rho_v(t)$ where the couplings between the electronic and nuclear degrees of freedom for the molecular system are included. The operator $V(t)$ is given by

$$V(t) = \sum_n (Y_n(t)B_n^\dagger + Y_n^*(t)B_n) \tag{19.35}$$

and the changes in the statistical density operator for the molecular vibrational modes $\rho_v(t)$ are given by the unitary transformation where the time-dependent coupling parameters $Y_n(t)$, $Y_n^*(t)$ determine how substantial the changes are.

19.4 THE EQUATION OF MOTION

Next we will consider the actual expressions in the equation of motions for the different operators. The evolution of the electronic operators $\{a_r^\dagger a_s, a_s^\dagger a_r\}$ depends on the expectation values of the operators $\{b_k^\dagger, b_k\}$ and $\{B_n^\dagger, B_n\}$ since

$$i \frac{d}{dt} \langle a_r^\dagger a_s \rangle = \langle [a_r^\dagger a_s, H_e + H_v + W_{e-v} + H_s + W_{o-s} + H_r + W_{o-r}] \rangle$$

$$= \langle [a_r^\dagger a_s, H_e] \rangle_e + \langle [a_r^\dagger a_s, W_{e-v}] \rangle_{e,\text{vib}}$$

$$+ \langle [a_r^\dagger a_s, W_{o-s}] \rangle_{o,s} + \langle [a_r^\dagger a_s, W_{o-r}] \rangle_{o,r}$$

$$= \langle [a_r^\dagger a_s, H_e] \rangle_e + \sum_n A_n \langle [a_r^\dagger a_s, H_n^{(1)}] \rangle_e (\langle B_n^\dagger \rangle_{\text{vib}} + \langle B_n \rangle_{\text{vib}})$$

$$+ \sum_k \alpha_k \langle [a_r^\dagger a_s, M_k] \rangle_e \langle b_k^\dagger \rangle_s$$

$$+ \sum_k \alpha_k^* \langle [a_r^\dagger a_s, M_k^\dagger] \rangle_e \langle b_k \rangle_s$$

$$+ \sum_{p\lambda} \sqrt{\frac{2\pi}{pcV}} \{ \langle [a_r^\dagger a_s, j_\lambda(\mathbf{p})] \rangle_e \langle d_{p\lambda}^\dagger \rangle_r + \langle [a_r^\dagger a_s, j_\lambda^\dagger(\mathbf{p})] \rangle_e \langle d_{p\lambda} \rangle_r \}$$

$$\tag{19.36}$$

The evolution of the operators for the outer medium $\{b_k^\dagger, b_k\}$, the intramolecular degrees of freedom $\{B_n^\dagger, B_n\}$ and the field operators is given by

$$i\frac{d}{dt}\langle b_k\rangle_s = \langle [b_k, H_s]\rangle_s + \langle [b_k, W_{o-s}]\rangle_{e,s}$$

$$= \langle [b_k, H_s]\rangle_s + \sum_{k'}\langle [b_k, b_{k'}^\dagger]\rangle_s\langle \alpha_{k'}M_{k'}\rangle_e$$

$$+ \sum_{k'}\langle [b_k, b_{k'}]\rangle_s\langle \alpha_{k'}^*M_{k'}^\dagger\rangle_e \qquad (19.37)$$

and

$$i\frac{d}{dt}\langle B_m\rangle_{\text{vib}} = \langle [B_m, H_v]\rangle_{\text{vib}} + \langle [B_m, W_{e-v}]\rangle_{e-\text{vib}}$$

$$= \langle [B_m, H_v]\rangle_{\text{vib}}$$

$$+ \sum_{m'}\langle [B_m, (B_{m'}^\dagger + B_{m'})]\rangle_{\text{vib}}\langle A_{m'}H_{m'}^{(1)}\rangle_e \qquad (19.38)$$

and finally

$$i\frac{d}{dt}\langle d_{p\lambda}\rangle_r = \langle [d_{p\lambda}, H_r]\rangle_r + \langle [d_{p\lambda}, W_{o-r}]\rangle_{e,r}$$

$$= \langle [d_{p\lambda}, H_s]\rangle_r + \sum_{p'\lambda'}\sqrt{\frac{2\pi}{p'cV}}\left(\langle [d_{p\lambda}, d_{p'\lambda'}^\dagger]\rangle_r\langle j_{\lambda'}(\mathbf{p'})\rangle_e\right.$$

$$\left. + \langle [d_{p\lambda}, d_{p'\lambda'}]\rangle_r\langle j_{\lambda'}^\dagger(\mathbf{p'})\rangle_e\right) \qquad (19.39)$$

The equations require calculation of the energies and coupling elements between the different subsystems. Having obtained those, one is able to perform the propagation of these equations and investigate the effects of the laser field on the charge transfer process.

19.5 DISCUSSION OF THE MODEL

If we are in the limit where we can neglect the reservoir modes, we obtain a picture of a molecular system dressed by laser photons. The term dressed denotes the coupling between the molecule and the laser field. The coupling of the reservoir of the initially empty photon modes to the dressed molecular system gives rise to radiative cascading to other dressed molecular states. Our description of the quantum description of the laser field is greatly simplified by considering a laser cavity (of volume V) possessing only one mode (of frequency ω_L) containing photons. In the case of a molecular system, which can be represented as a two-level molecule coupled to a laser field, we can transform from the states $|GS, N + 1\rangle$ and $|CT, N\rangle$ into dressed molecular states. These two states are:

1. The state defined by $|GS, N + 1\rangle$; the molecular ground state plus $N + 1$ photons
2. The state defined by $|CT, N\rangle$; the molecular charge transfer state plus N photons

During the $GS \rightarrow CT$ transition one photon is absorbed.

Next we will consider the situation where the frequency of the laser mode (ω_L) is detuned by an amount δ from the frequency of the molecular transition (ω)

$$\delta = \omega - \omega_L \tag{19.40}$$

and we obtain what is termed resonant or quasi-resonant laser fields. These laser fields give rise to substantial changes, where the two states $|GS, N + 1\rangle$ and $|CT, N\rangle$ are nearly degenerate and belong to a submanifold $P_N = \{|GS, N + 1\rangle, |CT, N\rangle\}$, the complete manifold being given by $\{P_N\}$ for all N. Within a given submanifold we perform a diagonalization of the energy expression in terms of the two states in the submanifold.

We observe that the eigenvalues change with the detuning of the laser field or the coupling strength between the molecular states and the laser mode. Fig. 19.1 presents schematically how the eigenvalues change according to changes in the detuning and the photon number, the latter reflecting the coupling strength. Coupling to the reservoir causes photons to appear in modes that initially were empty. These are the fluorescence photons that the excited molecular state emits.

In Fig. 19.2 we present the time evolution of the populations of the $|CT, N\rangle$ state. The molecular system is *para*-nitroaniline solvated by a dielectric medium having the dielectric properties of benzene. The external laser field has the following characteristics:

(1) The pulse length is 2150 a.u. or 52.01 fs
(2) The number of photons in the field is 80
(3) The time delay between pulses is 1000 a.u. or 24.19 fs

The dependence of the number of photons (the intensity of the field) on the excitation energies and transition moments were calculated using the methods presented in Chapter 15.

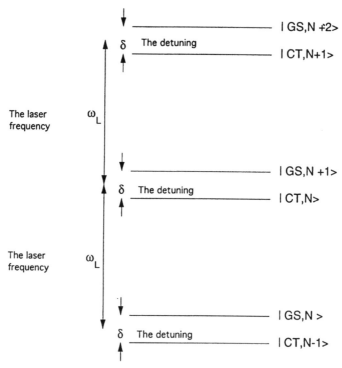

Fig. 19.1. A schematic energy diagram of the dressed molecule.

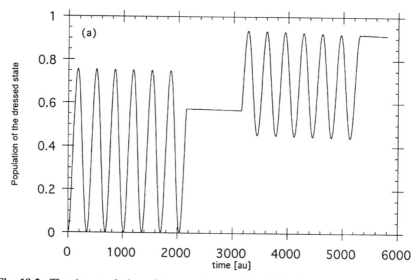

Fig. 19.2. The time evolution of the population of the $|CT, N\rangle$ state for three different line width of the CT state: (**a**) the line width of the CT state is 10^{-5} a.u. (**b**) the line width of the CT state is 10^{-4} a.u. (**c**) the line width of the CT state is 10^{-3} a.u.

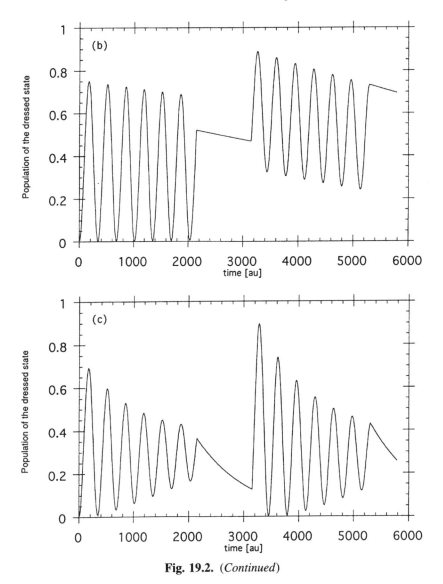

Fig. 19.2. (*Continued*)

The three different figures present the effect of the line width of the charge transfer (CT) state, all of them being larger than the natural line width of the state (see ref. [316]) for further details). We observe further a rather strong influence of damping on the populations of the $|CT, N\rangle$ state. The line width of the state has been increased in order to illustrate the effects of broadening mechanisms or losses.

20

PROTON TRANSFER REACTIONS IN SOLUTION

Historically there have been two approaches for describing proton transfer reactions in solution: one by Bell [318] and one by Marcus [319]. The model of Bell is based on transition state theory and includes quantum corrections taking account possible tunneling of the proton before reaching the transition state configuration. The model of Marcus focuses on the interactions between the reacting complex and the polar solvent during the charge transfer and is closely related to Marcus' picture of electron transfer reactions. In the present chapter we will outline some of the more recent work in this area [320–322].

We will consider a proton transfer reaction in an inter- or intramolecular complex solvated in a polar solvent. The reaction is pictured as being unimolecular within the complex, and the actual diffusion of the proton donor and acceptor for forming the intermolecular complex is neglected. The reaction is written as

$$(DH + A) \rightleftharpoons (D^- + HA^+) \tag{20.1}$$

where the ( ) signifies the existence of the molecular complex within which the proton transfer takes place. As a first approximation the reaction is characterized by two coordinates: (i) the position of the proton, given by r; and (ii) the distance between the proton donor (D) and acceptor (A) sites, given by R.

The inter- or intramolecular potential will for a given value of R be a double-well potential, where each minimum corresponds to the proton being localized either on the D-site or the A-site. The form of the potential and the size of the activation barrier are strongly dependent on the distance between the two sites, D and A. In general, one finds that for shorter D–A distances, the activation energy for proton transfer is lower than that for longer D–A distances. In order to form the configuration leading to a shorter D–A distance, one has to overcome the short-range repulsion between the two sites. The transfer of a proton leads to changes in the charge distribution of the molecular complex, and these changes will lead to different solvent interactions for the two different charge distributions, the charge distribution of the initial or final state of the reaction. The solvent interactions will change the potential energy surface and the dynamics of the reaction. The solvent response will, as in the case of electron transfer, depend on the timescale of the actual transfer of the proton. If the proton transfer occurs on a longer timescale than the relaxation of the inertial polarization of the solvent, then one can safely assume that the solvent is fully equilibrated along the reaction path. If the opposite is the case, then the relaxation processes of the solvent will influence the dynamics of the reaction, and one could imagine effects such as those discussed in Chapters 14 and 15 of ref. [317].

Traditionally proton transfer reactions are divided into two classes: (i) those involving weak H-bonds or the nonadiabatic limit, and (ii) those involving strong H-bonds or the adiabatic limit. The former class is depicted with a double-well potential energy surface having a large activation barrier between the reactant and product configurations. The latter class is assumed to have a small activation barrier or none at all, leading to a rather broad and flat potential energy surface. In both cases interactions between the molecular system and the solvent will lead to changes in the potential energy surface, which could be a more asymmetric potential energy surface illustrating the stabilizing effect of the polar solvent towards charged compounds.

20.1 THE HAMILTONIAN

The molecular complex undergoing a proton transfer reaction is surrounded by a solvent consisting of N polar molecules. We let the Hamiltonian for the total system be written as

$$H = H_{mol} + H_{sol} + V_{int} \tag{20.2}$$

where H_{mol} and H_{sol} are the Hamiltonians of the molecular complex and the solvent, respectively. They are given by

$$H_{mol} = T_r + T_R + T_Q + U(r, R, \mathbf{Q}) \tag{20.3}$$

and

$$H_{sol} = T_{sol} + U_{sol} \tag{20.4}$$

The potential $U(r, R, \mathbf{Q})$ depends on the previously discussed distances r and R, along with the rest of the internal coordinates $\mathbf{Q}$. We assume that the potential $U(r, R, \mathbf{Q})$ has been established from either gas phase calculations or semiempirical potentials. The kinetic energy operators T_r, T_R, T_Q are related to the kinetic energy of the proton, the relative distance between the two sites, and the internal motion of the molecular complex. The kinetic energy operator T_{sol} takes care of the kinetic energy of the solvent degrees of freedom, whereas U_{sol} represent the potential energy within the solvent. Interactions between the molecular complex and the solvent are given by V_{int} which may take the form

$$V_{int} = -\mathbf{d}_{mol} \cdot \mathbf{E}_{sol} \tag{20.5}$$

where $\mathbf{d}_{sol}$ is the dipole moment of the molecular complex and $\mathbf{E}_{sol}$ is the electric field from the solvent. This expression is rather problematic since it assumes a dipole approximation of the charge distribution of the molecular complex and also that the charge distribution is unaffected by the presence of the solvent. A general expression would be of the type encountered in Chapter 14: Nonequilibrium Solvation.

Evaluation of a rate constant requires a separation of the reactant and product region, which can be done by the operator [320–326]

$$N_{reac} = 1 - \Theta(r - r^*) \tag{20.6}$$

where $\Theta(r - r^*)$ is the Heaviside function and r^* is the transition state configuration for the potential $U(r, R, \mathbf{Q})$ with respect to r. The linear response expression by Yamamoto [323] gives the following rate coefficient, μ,

$$\mu = \int_0^t ds \left[\frac{2\pi}{\beta h} \int_0^{\beta h} d\tau < \dot{N}_{reac} \dot{N}_{reac}(s + i\tau) > \right] \tag{20.7}$$

The rate constants for the backward (b) and forward (f) reactions are

$$k_f = \frac{Q}{Q_R} \mu \tag{20.8}$$

$$k_b = \frac{Q}{Q_P} \mu \tag{20.9}$$

where

$$Q = \text{Tr}\{\exp(-\beta H)\} \tag{20.10}$$

$$Q_R = \text{Tr}\{N_{\text{reac}} \exp(-\beta H)\} \tag{20.11}$$

$$Q_P = Q - Q_R \tag{20.12}$$

The time derivative used in Eq. (20.7) is given as

$$\dot{N}_{\text{reac}} = -\frac{2i\pi}{h} [H, N_{\text{reac}}] \tag{20.13}$$

The upper time limit in Eq. (20.7) should be large compared to the timescales of the molecular motion.

In order to proceed we assume that the Born-Oppenheimer (BO) approximation is valid not only for the electronic and nuclei degrees of freedom but also for the motion of the transferable proton and the rest of the molecular modes.

We let the wave function for the system be written as

$$|\Psi\rangle = |\psi_e\rangle |\psi_p\rangle |\psi_{R,Q}\rangle |\psi_S\rangle \tag{20.14}$$

where

- $|\psi_e\rangle$ is the electronic wave function
- $|\psi_p\rangle$ is the wave function for the proton
- $|\psi_{R,Q}\rangle$ is the wave function for all the other molecular degrees of freedom
- $|\psi_S\rangle$ is the wave function for the solvent degrees of freedom

By applying the BO approximation between the electronic and nuclear degrees of freedom we obtain the effective Hamiltonian for the nuclear motion given in Eq. (20.3). The second level of the BO approximation gives an effective Hamiltonian for the solvent degrees of motion and the molecular modes.

$$\langle \psi_p | H | \psi_p \rangle = \langle \psi_p | T_r | \psi_p \rangle + T_R + T_Q + T_{\text{sol}}$$

$$+ \langle \psi_p | U(r, R, Q) | \psi_p \rangle + \langle \psi_p | V_{\text{int}} | \psi_p \rangle + U_{\text{sol}}$$

$$= T_R + T_{\text{sol}} + E_{\text{prot}, n}(R, Q, S) \tag{20.15}$$

where

$$E_{\text{prot}, n}(R, Q, S) = \langle \psi_p | T_r | \psi_p \rangle + \langle \psi_p | U(r, R, Q) | \psi_p \rangle + \langle \psi_p | V_{\text{int}} | \psi_p \rangle + U_{\text{sol}} \tag{20.16}$$

which is the nth eigenvalue for the nth proton eigenfunction.

We then have an effective potential for the molecular motion and the solvent motion. From the eigenvalue equation

$$E_{\text{prot},n}(R, \mathbf{Q}, S)|\psi_{\text{prot},n}\rangle = [T_r + U(r, R, \mathbf{Q}) + V_{\text{int}} + U_{\text{sol}}]|\psi_p\rangle = H_n|\psi_p\rangle \quad (20.17)$$

we obtain the eigenfunctions and eigenvalues for the motion of the proton. For simplicity we limit ourselves to the situation where the proton transfer reaction can be described by two eigenstates corresponding to Eq. (20.17); we denote them $|\psi_{\text{prot},0}\rangle$ and $|\psi_{\text{prot},1}\rangle$, the ground state and the first excited state, respectively.

The states $|\psi_{\text{prot},n}\rangle$ are eigenstates that are delocalized with respect to the double-well potential. In order to be able to determine whether the proton is localized on the D or A site we define the wave function [320–322]

$$|\Theta_{\text{reac}}\rangle = C_0|\psi_{\text{prot},0}\rangle + C_1|\psi_{\text{prot},1}\rangle \quad (20.18)$$
$$|\Theta_{\text{prod}}\rangle = C_1|\psi_{\text{prot},0}\rangle - C_0|\psi_{\text{prot},1}\rangle \quad (20.19)$$

where

$$C_0^2 = \int_{-\infty}^{r*} dr\,|\psi_{\text{prot},0}|^2 \quad (20.20)$$

and

$$C_1^2 = 1 - C_0^2 \quad (20.21)$$

These coefficients will be time-dependent in order to follow the proton transfer reaction. We are then able to rewrite the Hamiltonian to get [320–322]

$$H = |\Theta_{\text{reac}}\rangle H_{\text{reac}}(R, \mathbf{Q}, S)\langle\Theta_{\text{reac}}| + |\Theta_{\text{prod}}\rangle H_{\text{prod}}(R, \mathbf{Q}, S)\langle\Theta_{\text{prod}}|$$
$$+ \gamma(R, \mathbf{Q}, S)(|\Theta_{\text{reac}}\rangle\langle\Theta_{\text{prod}}| + |\Theta_{\text{prod}}\rangle\langle\Theta_{\text{reac}}|) \quad (20.22)$$

where

$$H_{\text{reac}}(R, \mathbf{Q}, S) = C_0^2 H_0 + C_1^2 H_1 \quad (20.23)$$
$$H_{\text{prod}}(R, \mathbf{Q}, S) = C_1^2 H_0 + C_0^2 H_1 \quad (20.24)$$
$$\gamma(R, \mathbf{Q}, S) = C_0 C_1(H_1 - H_0) \quad (20.25)$$

The operator N_{reac} can be specified as

$$N_{\text{reac}} = |\Theta_{\text{reac}}\rangle\langle\Theta_{\text{reac}}| \quad (20.26)$$

and the expectation value $\langle N_{\text{reac}}\rangle$ gives the probability of finding the proton on the donor site.

For proton transfer in a molecular complex having a large activation energy for all donor–acceptor site distances, we are in the weakly H-bonded group also termed the nonadiabatic limit since the proton coupling is small and therefore leads to a large activation energy. The proton transfer coupling is small compared to the energy difference related to

$$\Delta E = \langle \Theta_{prod} | H | \Theta_{prod} \rangle - \langle \Theta_{reac} | H | \Theta_{reac} \rangle \qquad (20.27)$$

and it is appropriate to make use of the wave functions $|\Theta_{reac}\rangle$ and $|\Theta_{prod}\rangle$ since these represent the localized states for the proton on the donor or acceptor site.

From Eqs. (20.13) and (20.26) we obtain [320–322]

$$\dot{N}_{reac} = -\frac{2i\pi}{h}\, \gamma(R, Q, S)(|\Theta_{reac}\rangle\langle\Theta_{prod}| + |\Theta_{prod}\rangle\langle\Theta_{reac}|) \qquad (20.28)$$

and using this expression in Eq. (20.7) we get as the dominant term with respect to γ

$$\mu = \frac{4\pi^2 Q_R}{Qh^2} \int_{-\infty}^{\infty} dt \left\langle \gamma(t = 0)\exp\left[\frac{2i\pi}{h}\int_0^t d\tau \Delta E(\tau)\right] \gamma(t)\right\rangle_R \qquad (20.29)$$

where the expectation values and operators are defined in terms of

$$\langle \Lambda \rangle = \frac{1}{Q_R}\, \mathrm{Tr}\{\Lambda \exp(-\beta H)\} \qquad (20.30)$$

where

$$\Lambda = \exp\left(\frac{2i\pi H_R t}{h}\right) \Lambda \exp\left(-\frac{2i\pi H_R t}{h}\right) \qquad (20.31)$$

When treating the motion of R and Q quantum mechanically, it is important to keep the time-ordering. The time-dependent energy difference represents the time-dependent splitting between the two diabatic states: (i) the proton being on the donor; (ii) the proton being on the acceptor. This energy difference will depend on the evolution of the molecular modes (R, Q) and the solvent modes. The time-dependent coupling γ will depend on the barrier height, which is strongly influenced by the distance between the donor and acceptor sites but also by the presence of the remaining molecular modes along with the solvent modes. Let us, however, make the assumption that γ only depends on R, and let the dependence be written as

$$\gamma(R) = \gamma_o \exp(-\alpha(R - R_o) - \omega(R - R_o)^2) \qquad (20.32)$$

where R_o is an appropriate chosen expansion point, usually at the transition state configuration, and γ_o is the coupling at that configuration [320–322]. Utilizing this assumption for γ we can rewrite the expression for the rate constant as

$$
\mu = \frac{8\pi^2\gamma_o^2}{h^2} \int_0^\infty dt\, \exp\left[\frac{2i\pi}{h} \langle\Delta E\rangle t \right.
$$

$$
- \frac{4\pi^2}{h^2} \int_0^t \int_0^{\tau'} d\tau' \langle \delta E \delta E(\tau') \rangle
$$

$$
\left. + \alpha^2 (\langle \delta R \delta R(t)\rangle + \langle \delta R^2 \rangle) \right] \tag{20.33}
$$

where

$$
\delta E = \Delta E - \langle\Delta E\rangle \tag{20.34}
$$

and

$$
\delta R = R - R_o \tag{20.35}
$$

Our problem in this case has been reduced to finding the correlation functions $\langle \delta E \delta E(\tau') \rangle$ and $\langle \delta R \delta R(t) \rangle$.

In the case of proton transfer along strongly H-bonded systems, we have to approach the problem in a slightly different manner. In this case, the difference between the energies related to the two adiabatic Hamiltonians, H_0 and H_1 is large compared to kT, and it is safe to assume that the dynamics of the reaction take place on the lowest adiabatic surface (H_0). The expectation value of any quantum operator is to be performed with respect to the ground state wave function for the proton motion: $|\psi_{\text{prot},0}\rangle$. We then have

$$
\langle \psi_{\text{prot},0} | N_{\text{reac}} | \psi_{\text{prot},0} \rangle = C_0^2 \tag{20.36}
$$

where C_0^2 is the weight of the ground state wave function for having the proton on the donor site. The rate coefficient μ is given by

$$
\mu = \frac{1}{2} \int_{-\infty}^\infty dt \langle \dot{C}_0^2 \dot{C}_0^2(t) \rangle \tag{20.37}
$$

where C_0 is 1 in the reactant side and 0 in the product side; C_0 is found to vary strongly in the phase space region related to the area where $\Delta E \approx 0$, the transition state area [320–322].

APPENDIX A

Consider the equation

$$i\hbar \dot{U}(t, t_0) = A(t)U(t, t_0) \qquad (A.1)$$

It has been shown (see Ref. [4]) that the solution can be expressed as

$$
U(t, t_0) = \exp\left(\int_{t_0}^{t} dt' \, \frac{A(t')}{i\hbar} + \frac{1}{2} \int_{t_0}^{t} dt' \int_{t_0}^{t'} dt'' \left[\frac{A(t')}{i\hbar}, \frac{A(t'')}{i\hbar} \right] \right.
$$

$$
+ \frac{1}{6} \int_{t_0}^{t} dt' \int_{t_0}^{t'} dt'' \int_{t_0}^{t''} dt''' \left\{ \left[\frac{A(t''')}{i\hbar}, \left[\frac{A(t'')}{i\hbar}, \frac{A(t')}{i\hbar} \right] \right] \right.
$$

$$
\left. \left. + \left[\left[\frac{A(t''')}{i\hbar}, \frac{A(t'')}{i\hbar} \right], \frac{A(t')}{i\hbar} \right] \right\} + \dots \right) \qquad (A.2)
$$

We notice that the higher order terms of this infinite series are zero whenever a commutator of a lower order is zero. This happens in the case of the linearly forced harmonic oscillator, where we have

$$A(t) = p(t)a^{\dagger} + q(t)a \qquad (A.3)$$

Thus the commutator

$$[A(t'), A(t'')] = p(t')q(t'')[a^{\dagger}, a] + q(t')p(t'')[a, a^{\dagger}] = q(t')p(t'') - p(t')q(t'') \qquad (A.4)$$

is just a function of t' and t''. It will therefore commute with $A(t)$ and hence the series truncates after the second term.

APPENDIX B

In Section 2.1.4 we considered the solution to the set of equations

$$i\hbar\dot{\mathbf{R}}(t, t_0) = \mathbf{B}\mathbf{R}(t, t_0) \tag{B.1}$$

where $\mathbf{R}(t_0, t_0) = \mathbf{I}$ the unit matrix. In order to obtain the amplitude for a transition from an initial state vector $|n_1 \ldots n_M\rangle$ to $|n_1' \ldots n_M'\rangle$ using Eq. (2.143), we need to evaluate the functions α_{ji}. We introduce the new functions

$$f_{ij} = \alpha_{ij} \, (i \neq j) \tag{B.2}$$
$$f_{ii} = \exp(\alpha_{ii}) - 1 \tag{B.3}$$

and obtain the functions f_{ij} in the following manner [7, 9]: For $m < n$ we have

$$f_{mn} = |\mathbf{A}|/|\mathbf{D}| \tag{B.4}$$

where $|\mathbf{A}|$ indicates the determinant of the matrix $\mathbf{A}$. The matrices $\mathbf{A}$ and $\mathbf{B}$ are of order $(n-1) \times (n-1)$ and the elements of the two matrices given as

$$D_{kl} = Q_{kl} \, (k \neq l) \tag{B.5}$$
$$D_{kk} = 1 + Q_{kk} \tag{B.6}$$

and

$$A_{km} = Q_{kn} \tag{B.7}$$
$$A_{kl} = D_{kl} \, (l \neq m) \tag{B.8}$$

where k and l runs from 1 to $n - 1$. For the element f_{nn} we have

$$f_{nn} = Q_{nn} - \sum_{m=1}^{n-1} f_{mn} Q_{nm} \qquad (B.9)$$

where the elements f_{mn} have just been determined. For f_{nm} where $m < n$ we obtain

$$f_{nm} = \frac{Q_{nm} - \sum_{k=1}^{m-1} Q_{nk} f_{km}}{1 + f_{nn}} \qquad (B.10)$$

Thus the above equations enable us to determine all the elements f_{nm} from the solution to the differential equation, Eq. (22.1), using the relation $\mathbf{Q} = \mathbf{R} - \mathbf{I}$. Instead of using the above equations, we may also use an iterative scheme (see ref. [9]).

APPENDIX C

ANSWERS TO EXERCISES

Chapter 2

2.1. The solution follows by use of the commutation relations.

2.2. The result is $-|1, 1, 0, \ldots\rangle$.

2.3. From the commutation relation for bosons we get $H_0 a^\dagger = a^\dagger (H_0 + \hbar\omega)$, i.e., $H_0^n a^\dagger = a^\dagger (H_0 + \hbar\omega)^n$, from which the result follows by summing the Taylor series.

2.4. This follows from $\Delta E = \sum_{n=0}(E_n - E_0)P_{n0} = \hbar\omega \exp(-\rho) \sum_{n=0} n\rho^n/n! = \hbar\omega\rho$. It may also be shown that this result is independent of the initial state of the oscillator.

2.5. The solution to the classical equations of motion can be found as:

$$Q(t) = a \cos(\omega t + \phi) - \frac{1}{\omega} \int_{t_0}^{t} dt' f(t') \sin[\omega(t - t')] \qquad (C.1)$$

where $a = \sqrt{E_0}/\omega$, E_0 the initial energy of the oscillator, and ϕ a phase angle.

2.6. Use the relationships $b \exp(\alpha^- b^\dagger) = \exp(\alpha^- b^\dagger)(b + \alpha^-)$ and $b\Phi_0 = 0$.

2.7. First we notice that

$$\Phi_- = \exp(-\rho/2) \exp(\alpha^- b^\dagger)\Phi_0 \qquad (C.2)$$

By using the commutation relation for bosons we may show that

$$b \exp(\alpha^- b^\dagger) = \exp(\alpha^- b^\dagger)(b + \alpha^-) \qquad (C.3)$$

Thus we have

$$(\hbar\omega b^\dagger b + \gamma^* b^\dagger + \gamma b) \exp(\alpha^- b^\dagger)\Phi_0$$
$$= \exp(\alpha^- b^\dagger)[\hbar\omega(b^\dagger b + \alpha^- b^\dagger) + \gamma^* b^\dagger + \gamma(\alpha^- + b)]\Phi_0 \qquad (C.4)$$

Since $a\Phi_0 = 0$, we see that Φ_- is an eigenfunction if $\hbar\omega\alpha^- + \gamma^* = 0$ with the eigenvalue $\gamma\alpha^-$. Thus the eigenvalue is $-|\gamma|^2/\hbar\omega$.

2.8. We use the relation

$$\frac{\partial H_{\text{eff}}}{\partial t} = \langle\dot{\psi}|H|\psi\rangle + \langle\psi|H|\dot{\psi}\rangle \qquad (C.5)$$

and from $i\hbar\dot{\psi} = H_q\psi$ we get the desired result, since H and H_q commute.

2.9. We obtain

$$\dot{\rho}_k = \frac{1}{\hbar\omega_k} V_k^{(1)} \int dt'\, V_k^{(1)} \cos[\omega_k(t - t')]$$

and

$$\frac{\partial\epsilon_k}{\partial t} = -V_k^{(1)} \int dt'\, V_k^{(1)} \cos[\omega_k(t - t')]$$

from which we obtain the desired result.

Chapter 3

3.1. The normalization condition gives

$$\int dx\, \phi^*\phi = \int dx \exp\{-2\mathrm{Im}\,\gamma/\hbar - 2\mathrm{Im}\,A[x - x(t)]^2/\hbar\}$$
$$= \exp(-2\mathrm{Im}\gamma/\hbar)\sqrt{\pi\hbar/2\mathrm{Im}\gamma} = 1 \qquad (C.6)$$

3.2. We get $x(t) = x_0 \cos(\omega t)$, $p(t) = -\omega x_0 \sin(\omega t)$, and

$$A(t) = a \; \frac{iA(0) \cos(\omega t) - a \sin(\omega t)}{iA(0) \cos(\omega t) + a \cos(\omega t)} \tag{C.7}$$

If we insert this expression in the equation for dA/dt and use that the force constant is $k = m\omega^2$, we get $a = m\omega/2$. A minimum uncertainty wavepacket is obtained if $A(0) = a$. This wavepacket will keep its width in time and be centered around the classical trajectory $x(t)$. In general we have

$$|\Psi(x, t)|^2 = \frac{1}{\sqrt{\pi w(t)}} \exp\left[-\frac{1}{w(t)} (x - x(t))^2 \right] \tag{C.8}$$

where

$$w(t) = \frac{\hbar}{m\omega} \; \frac{a^2 \cos^2(\omega t) + A(0)^2 \sin^2(\omega t)}{aA(0)} \tag{C.9}$$

and

$$\text{Re}A(t) = \frac{m\omega}{2} \; \frac{(A(0)^2 - a^2) \cos(\omega t) \sin(\omega t)}{a^2 \cos^2(\omega t) + A(0)^2 \sin^2(\omega t)} \tag{C.10}$$

3.3. Inserting $UP_0 = (P_0 + X)\tilde{U}$ in the TDSE we get

$$i\hbar \; \frac{\partial U}{\partial t} \; P_0 = i\hbar \; \frac{\partial X}{\partial t} \; \tilde{U} + i\hbar(P_0 + X) \; \frac{\partial \tilde{U}}{\partial t} \tag{C.11}$$

which gives

$$HUP_0 = i\hbar \; \frac{\partial X}{\partial t} \; \tilde{U} + (P_0 + X)H_{\text{eff}}\tilde{U} \tag{C.12}$$

Using that $UP_0 = \Omega \tilde{U}$ and $\Omega = P_0 + X$ we get

$$i\hbar \; \frac{\partial X}{\partial t} = (Q_0 - XP_0)H(P_0 + X) \tag{C.13}$$

from which the result is obtained by using $XP_0 = X = Q_0X$.

Chapter 4

4.1. We expand the phase as

$$S(x, x_1) = S(x_2, x_1) + \left.\frac{\partial S}{\partial x_2}\right|_x (x_2 - x)$$

i.e., we have

$$\int dx_1 |F|^2 \exp\left[\frac{i}{\hbar}\left.\frac{\partial S}{\partial x_2}\right|_x (x_2 - x)\right] = \delta(x - x_2) \qquad (C.14)$$

Thus introducing

$$y = x - x_2, \ z = \frac{1}{\hbar}\left.\frac{\partial S}{\partial x_2}\right|_x$$

and

$$dx_1 = dz \left|\frac{\partial x_1}{\partial z}\right|$$

we get the desired result.

4.2. Using that

$$S_{cl} = \int_{t_1}^{t_2} dt \, \frac{m}{2} \dot{x}^2 = \frac{m}{2}\left(x\dot{x}|_1^2 - \int_{t_1}^{t_2} dt \, x\ddot{x}\right) \qquad (C.15)$$

we get for a free particle that

$$S_{cl} = \frac{m}{2} \frac{(x_2 - x_1)^2}{t_2 - t_1} \qquad (C.16)$$

where we have used that $\ddot{x} = 0$ and $\dot{x} = (x_2 - x_1)/(t_2 - t_1)$.
For the harmonic oscillator we obtain the solution by expressing $x(t)$ as

$$x(t) = a\cos(\omega t) + b\sin(\omega t) \qquad (C.17)$$

where the boundary conditions give

$$a = \frac{x_1 \sin(\omega t_2) - x_2 \cos(\omega t_1)}{\sin(\omega(t_2 - t_1))} \tag{C.18}$$

$$b = \frac{x_2 \cos(\omega t_1) - x_1 \cos(\omega t_2)}{\sin[\omega(t_2 - t_1)]} \tag{C.19}$$

which yields

$$S_{cl} = \frac{m\omega}{2 \sin[\omega(t_2 - t_1)]} \{(x_1^2 + x_2^2) \cos[\omega(t_2 - t_1)] - 2x_1 x_2\} \tag{C.20}$$

For the free particle we then have

$$\frac{\partial^2 S_{cl}}{\partial x_1 \partial x_2} = \frac{-im}{\hbar(t_2 - t_1)}$$

and for the harmonic oscillator

$$\frac{\partial^2 S_{cl}}{\partial x_1 \partial x_2} = \frac{-im\omega}{\sin[\omega(t_2 - t_1)]}$$

4.3. For a harmonic oscillator we have

$$\frac{p_r^2}{2m} + \frac{1}{2} kr^2 = E = \frac{1}{2} kr_0^2 \tag{C.21}$$

where $\pm r_0$ are the classical turning points and $k = m\omega^2$. Thus we get

$$n = \frac{m\omega}{\pi} \int_{-r_0}^{r_0} dr \sqrt{r_0^2 - r^2} = \frac{m\omega r_0^2}{2\pi} \arcsin(r/r_0)|_{-r_0}^{r_0} = \frac{E}{\omega} \tag{C.22}$$

Thus $E = n\omega$. Furthermore, we have from $p_r = \partial F_2/\partial r$

$$\frac{1}{2m} \left(\frac{\partial F_2}{\partial r}\right)^2 + \frac{1}{2} kr^2 = E \tag{C.23}$$

or

$$F_2 = \int dr \sqrt{2m\left(E - \frac{1}{2}kr^2\right)} \qquad (C.24)$$

and hence

$$\phi = \frac{\partial F_2}{\partial n} = \sqrt{m/2}\ \frac{\partial E}{\partial n} \int dr\ \frac{1}{\sqrt{E - kr^2/2}}$$

$$= \omega\sqrt{m/k} \int dr\,(r_0^2 - r^2)^{-1/2} = \arccos(r/r_0) \qquad (C.25)$$

or $r = r_0 \cos\phi$.

Chapter 5

5.1. The Baker-Hausdorff formula gives $\exp(A)\exp(B) = \exp(C)$ where $C = A + B + \frac{1}{2}[A, B] + O(\Delta t^3)$. Thus $\exp(B)\exp(C) = \exp(A + 2B + \frac{1}{2}([A, B] + [B, C]) + O(\Delta t^3)$. Since $[A, B] + [B, C] = O(\Delta t^3)$ the method is a second order method.

5.2. Using the propagator we get

$$\psi(x, t) = \int dx_0\ K(x, t; x_0, t_0)\psi(x_0, t_0) \qquad (C.29)$$

Thus we have

$$\psi_E^+ = \frac{1}{\sqrt{2\pi}} \int_0^\infty dt \exp(iEt/\hbar) \int dx_0\ K(x, t; x_0, t_0)\psi(x_0, t_0) \qquad (C.30)$$

With the free-particle propagator and introducing $t_0 = 0$ and $u^2 = t - t_0$ we get

$$\psi_E^+ = \frac{2}{\sqrt{2\pi}} \sqrt{\frac{m}{2\pi i\hbar}} \int dx_0 \int du \exp[iEu^2/\hbar + im(x - x_0)^2/(2\hbar u^2)]\psi(x_0)$$

$$= \frac{m}{\hbar k}\exp(-ikx)\,\frac{1}{\sqrt{2\pi}} \int dx_0 \exp(ikx_0)\psi(x_0)$$

$$= \frac{m}{\hbar k}\exp(-ikx)\psi_k^+ \qquad (C.31)$$

where we have used the integral of Eq. (5.79) and $E = \hbar^2 k^2/2m$.

5.3. Using that

$$\int_{-\infty}^{\infty} dx \, \frac{\exp(-\beta x)}{1 + \exp(-qx)} = \frac{\pi}{q} \, \mathrm{cosec}(q\pi/q) \tag{C.32}$$

gives

$$k(T) = \frac{kT}{hQ(T)} \exp(-\beta V_0) \frac{\hbar\omega}{2kT} \, \mathrm{cosec}(\hbar\omega/2kT) \tag{C.33}$$

5.4. From the expression for the free particle propagator

$$\psi(x,t) = \int_{-\infty}^{\infty} dy \, K(x,t;y,t_0)\psi(y,t_0) \tag{C.34}$$

and

$$\int_{-\infty}^{\infty} dx \exp(-p^2 x^2 \pm qx) = \exp[q^2/(4p^2)]\sqrt{\pi}/p \tag{C.35}$$

we get

$$\psi(x,t) \sim \exp[-(x - x_{\mathrm{foc}})^2/4\Delta x^2 - ik_0 x] \tag{C.36}$$

Chapter 8

8.1. $2\pi E/\omega$.

8.2. a. $5.0 \cdot 10^{-15}$ cm^3/sec. **b.** $3.9 \cdot 10^{-15}$ cm^3/sec. **c.** 1.18 at 300 K. **d.** Using the transition state expression

$$k_{\mathrm{uni}}^{\infty} = \frac{kT}{h} \exp(-V_0/kT) \frac{\prod \omega_i}{\prod \omega_i^*} \frac{\sqrt{I_a^* I_b^* I_c^*}}{\sqrt{I_a I_b I_c}}$$

where asterisks (*) indicate the value at ($s = 0$), we get at 300 K, $k_{\mathrm{uni}}^{\infty} = 8.510^9 \cdot$ sec^{-1}.

8.3. Using that $R_- = 0$ and $L_+ = 1$ we get

$$L_- = \frac{-i\kappa \exp(-i\phi)}{\sqrt{1 + \kappa^2}} \tag{C.37}$$

$$R_+ = \frac{\exp(-i\phi)}{\sqrt{1 + \kappa^2}} \tag{C.38}$$

Thus the transmission factor is $T = 1/(1 + \kappa^2)$

8.4. By introducing the variable

$$x = \frac{2\pi}{\hbar\omega^*} (E - V_0)$$

we get

$$k(T) = \mu \exp(-\beta V_0) \int_{-x_0}^{\infty} dx \exp(-\mu x)/[1 + \exp(-x)] \tag{C.39}$$

where $\mu = \hbar\omega^*/(2\pi k T)$ and $x_0 = 2\pi V_0/(\hbar\omega^*)$. In the limit of $x_0 \rightarrow \infty$ we have

$$k(T) = \mu \exp(-\beta V_0) \operatorname{cosec}(\mu) \tag{C.40}$$

The Wigner tunneling correction is obtained using

$$\frac{\mu}{\sin \mu} \sim 1 + \frac{1}{6} \mu^2 \tag{C.41}$$

8.5. We introduce $x^2 = p^2/(2mkT)$ and a new function by $du = (\text{const})$ $\exp(-p^2/2mkT)\, dp$ or $u(x) = \operatorname{erf}(x)$, which gives $p = \pm\sqrt{2mkTx}$, with $\operatorname{erf}(x) = \xi$, where ξ is a random number between 0 and 1.

Chapter 9

9.1. Use that

$$\int_0^{\infty} ds \int_{-\infty}^{0} ds' \exp[-a(s^2 + s'^2) - 2bss']$$

$$= \frac{1}{2} \sqrt{\frac{\pi}{a}} \int_0^{\infty} ds' \exp[as'^2(b^2/a^2 - 1)][1 - \Phi(-bs'/\sqrt{a})]$$

$$= \frac{1}{2|b|} F\left[\frac{1}{2}, 1; \frac{3}{2}; (1 - b^2/a^2)\right] \tag{C.42}$$

where $b < 0$. Furthermore $b^2/a^2 < 1$.
Using

$$F\left[\frac{1}{2}, 1; \frac{3}{2}; (1 - b^2/a^2)\right] = \frac{1}{2z} \ln\left(\frac{1+z}{1-z}\right) \tag{C.43}$$

where

$$z = \sqrt{1 - a^2/b^2} = \frac{\sqrt{ch^2(\omega t) - \cos^2(0.5\hbar\omega\beta)}}{ch(\omega t)} \tag{C.44}$$

we get

$$C_{ss} = \frac{1}{4\pi} \frac{\sqrt{sh(\omega t_c)sh(\omega t_c^*)}}{ch(\omega t)} \frac{\exp(-\beta V_0)}{\sin(0.5\hbar\omega\beta)} F\left[\frac{1}{2}, 1; \frac{3}{2}; (1 - b^2/a^2)\right]$$

$$= \frac{kT}{h} \frac{0.5\hbar\omega\beta}{\sin(0.5\hbar\omega\beta)} \exp(-\beta V_0) f(t) \tag{C.45}$$

where

$$f(t) = \frac{1}{\omega} F\left[\frac{1}{2}, 1; \frac{3}{2}; (1 - b^2/a^2)\right] \frac{\sqrt{sh(\omega t_c)sh(\omega t_c^*)}}{ch(\omega t)} \tag{C.46}$$

which for $t \to \infty$ is

$$f(t) \to \frac{1}{\omega} \ln\left[\frac{ch(\omega t)}{\cos(0.5\hbar\omega\beta)}\right] \tag{C.47}$$

from which $\lim_{t\to\infty}(df(t)/dt) = 1$. Note that

$$a = \frac{m\omega}{2\hbar} \frac{\sin(\hbar\omega\beta)}{sh(\omega t_c)sh(\omega t_c^*)} \tag{C.48}$$

$$b = \frac{m\omega}{\hbar} \frac{ch(\omega t)\sin(0.5\hbar\omega\beta)}{sh(\omega t_c)sh(\omega t_c^*)} \tag{C.49}$$

$$sh(\omega t_c)sh(\omega t_c^*) = 0.5[ch(2\omega t) - \cos(\beta\hbar\omega)] \tag{C.50}$$

9.2. Use the propagator for the free particle in Eq. (4.14) or take the limit $\omega \to 0$ and $V_0 = 0$ in the expression for the parabolic barrier.

9.3. We obtain

$$k(T) = \frac{1}{Q(T)} \sum_i \exp(-\beta E_i) k_i \omega / (k_i + \omega) \tag{C.51}$$

where $Q(T)$ is the partition function for the $A + B$ system (per unit volume). The flux–flux correlation function corresponding to this expression is

$$C_{ff}(t) = \sum_i \exp(-\beta E_i)[k_i \omega \delta(t) - k_i^2 \exp(-k_i t)] \tag{C.52}$$

We have used that $\delta(t) = (d/dt)h(t)$ where $h(t)$ is the step function.

Chapter 10

10.1. From

$$d = \sum_{n=0} \sum_{j=0} (2j + 1)\sqrt{E - E_{\text{vib}} - E_{\text{rot}}} \tag{C.53}$$

we get

$$d = \frac{E^{5/2}}{\hbar \omega B_r} \int_0^1 df_R \int_0^{1-f_R} df_v \sqrt{1 - f_R - f_v} \tag{C.54}$$

where $B_r \doteq \hbar^2/2I$. Evaluating the integral yields: $d \sim 4E^{5/2}/(15\hbar\omega B_r)$ and finally

$$P_0 = \frac{15}{2} \frac{\hbar\omega\sqrt{B_r}}{E^{3/2}} \sqrt{f_R(1 - f_R - f_v)} \tag{C.55}$$

10.2. By using that

$$\int_0^P \sqrt{p - x} = -\frac{2}{3}(p - x)^{3/2}\Big|_0^P = \frac{2}{3} p^{3/2} \tag{C.56}$$

etc., we can evaluate the integral to be

$$\frac{2^M}{(2M+1)!!} \int_0^1 df_R \sqrt{f_R}(1-f_R)^{(2M+1)/2}$$

$$= \frac{2^M}{(2M+1)!!} \frac{\Gamma(3/2)\Gamma(M+3/2)}{\Gamma(M+3)} = \frac{\pi}{4(M+2)!} \qquad \text{(C.57)}$$

which finally gives the expression in Eq. 10.23.

10.3. This follows directly from the symmetry of the expression in Eq. 10.15.

10.4. By inserting

$$P(m,t) = P(n,t) + \frac{\partial P}{\partial n}(m-n) + \frac{1}{2}\frac{\partial^2 P}{\partial n^2}(m-n)^2 \qquad \text{(C.58)}$$

in Eq. (10.37) we get

$$B_0 = 0 \qquad \text{(C.59)}$$

$$B_1 = \sum_m A(n,m,t)(m-n) \qquad \text{(C.60)}$$

$$B_2 = \frac{1}{2}\sum_m A(n,m,t)(n-m)^2 \qquad \text{(C.61)}$$

10.5. Using the substitution $\exp(-\lambda\eta(t))u(x)$, with $x = n + \frac{1}{2}$ and $\eta(t) = \int^t dt' D(t')$, we obtain

$$xu''(x) + u'(x) + \lambda u(x) = 0 \qquad \text{(C.62)}$$

which has the solution $J_0(2(\lambda x)^{1/2})$. Thus the general solution is a linear combination of these solutions, i.e.,

$$P(x,t) = \int_0^\infty d\lambda a(\lambda) J_0(2\sqrt{\lambda x}) \qquad \text{(C.63)}$$

where the expansion coefficient $a(\lambda)$ can be obtained from the boundary condition, i.e., assuming $P(x,t_0) = \delta(x - x_0)$ we have

$$\delta(x - x_0) = \int_0^\infty d\lambda a(\lambda) J_0(2\sqrt{\lambda x}) \tag{C.64}$$

From Hankel's integral formula [82] we have

$$a(\lambda) = J_0(2\sqrt{\lambda x_0}) \tag{C.65}$$

which finally gives

$$P(x, t) = \int_0^\infty d\lambda J_0(2\sqrt{\lambda x_0}) J_0(2\sqrt{\lambda x}) \exp(-\lambda \eta(t))$$

$$= \frac{1}{\eta(t)} \exp\left[-\frac{1}{\eta(t)}(x + x_0)\right] I_0\left(\frac{2}{\eta(t)}\sqrt{xx_0}\right) \tag{C.66}$$

For small values of $\eta(t)$ we obtain

$$P(n, t) \sim \frac{1}{\eta} \exp\left[-\frac{1}{\eta}(\sqrt{n + 1/2} - \sqrt{n_0 + 1/2})^2\right] \tag{C.67}$$

Chapter 11

11.1. Consider the evaluation of the integral

$$\int_{-x_0}^{x_0} dx\, \delta(x) f(x) \tag{C.68}$$

where

$$\delta(x) = \frac{2}{\pi} \lim_{t \to \infty} \frac{\sin^2(xt/2)}{x^2 t} \tag{C.69}$$

Introducing the variable $y - xt$ we get

$$\lim_{t \to \infty} \frac{2}{\pi} \int_{-y_0}^{y_0} dy\, \frac{\sin^2(y/2)}{y^2} f(y/t)$$

$$= f(0) \frac{2}{\pi} \int_{-\infty}^{\infty} dy\, \frac{\sin^2(y/2)}{y^2} = f(0) \tag{C.70}$$

where we have used from ref. [83] the expression

$$\int_0^\infty \frac{\sin^2(ax)}{x^2}\, dx = \frac{a\pi}{2} \qquad\qquad \text{(C.71)}$$

11.2. Direct solution gives

$$G_{00} = \frac{z - H_{11}}{(z - H_{00})(z - H_{11}) - H_{01}H_{10}} \qquad\qquad \text{(C.72)}$$

We see that this is the fraction between the two determinants.

BIBLIOGRAPHY

[1] J. N. Murrell, S. Carter, S. C. Farantos, P. Huxley and A. J. C. Varandas, "Molecular Potential Energy Functions," John Wiley & Sons, New York, 1984.

[2] D. M. Hirst, "Potential Energy Surfaces," Taylor & Francis, London, 1985.

[3] P. A. M. Dirac, "The Principle of Quantum Mechanics," Clarendon Press, Oxford, 1958.

[4] W. Magnus, Commun. Pure Appl. Math. **7,** 649 (1954).

[5] R. D. Levine and C. E. Wulfman, Chem. Phys. Lett. **60,** 372 (1979); G. D. Billing and G. Jolicard, Chem. Phys. Lett. **102,** 491 (1983); C. J. Ballhausen, J. Phys. Chem. **99,** 2530 (1995).

[6] L. Infeld and T. E. Hull, Rev. Mod. Phys. **23,** 21 (1951).

[7] G. D. Billing, Comp. Phys. Rep. **1,** 237 (1984).

[8] G. D. Billing, Chem. Phys. **70,** 223 (1982).

[9] G. D. Billing, Chem. Phys. **51,** 417 (1980).

[10] G. D. Billing, Chem. Phys. **127,** 107 (1988); **104,** (1986).

[11] W. Miller, Jr., "Lie Theory and Special Functions," Academic Press, New York, 1968; J. Wei and E. Norman, J. Math. Phys. **4,** 575 (1963).

[12] G. D. Billing, J. Phys. Chem. **99,** 15378 (1995).

[13] R. B. Gerber, V. Buch and M. A. Ratner, J. Chem. Phys. **77,** 3022 (1982); R. B. Gerber and M. A. Ratner, J. Phys. Chem. **92,** 3252 (1988); Z. Kotler, A. Nitzan and R. Kosloff, Chem. Phys. Lett. **153,** 483 (1988); R. B. Gerber, R. Kosloff and M. Berman, Comp. Phys. Rep. **5,** 59 (1986).

[14] P. A. M. Dirac, Proc. Cambridge Phil. Soc. **26,** 376 (1930); J. Frenkel, "Wave Mechanics," Oxford University Press, Oxford, 1934.

[15] E. J. Heller, J. Chem. Phys. **62,** 1544 (1975).

[16] E. J. Heller, J. Chem. Phys. **64,** 63 (1976).

[17] G. D. Billing, Int. Revs. Phys. Chem. **13,** 309 (1994).

[18] H.-D. Meyer, U. Manthe and L. S. Cederbaum, Chem. Phys. Lett. **165,** 73 (1990); U. Manthe, H.-D. Meyer and L. S. Cederbaum, J. Chem. Phys. **97,** 9062 (1992); A. Jäckle and H.-D. Meyer, J. Chem. Phys. **102,** 5605 (1995).

[19] N. Makri and W. H. Miller, J. Chem. Phys. **87,** 5781 (1987); A. Garcia-Vela and R. B. Gerber, J. Chem. Phys. **98,** 427 (1993).

[20] G. D. Billing, Chem. Phys. **189,** 523 (1994).

[21] G. D. Billing and G. Jolicard, Chem. Phys. Lett. **221,** 75 (1994).

[22] G. D. Billing, Comp. Phys. Rep. **12,** 383 (1990); J. Phys. Chem. **99,** 15378 (1995).

[23] G. D. Billing, Chem. Phys. **70,** 223 (1982); **74,** 143 (1983).

[24] R. B. Gerber and R. Alimi, Chem. Phys. Lett. **184,** 69 (1991); N. Makri, Chem. Phys. Lett. **169,** 541 (1990); J. Campos-Martinez, J. R. Waldeck and R. D. Coalson, J. Chem. Phys. **96,** 3613 (1992).

[25] C. Bloch, Nucl. Phys. **6,** 329 (1958).

[26] G. Jolicard, Ann. Rev. Phys. Chem. **46,** 83 (1995).

[27] G. Jolicard and G. D. Billing, J. Chem. Phys. **101,** 9429 (1994).

[28] R. P. Feynman and A. R. Hibbs, "Quantum Mechanics and Path Integrals," McGraw-Hill, New York, 1965.

[29] M. F. Trotter, Proc. Am. Math. Soc. **10,** 545 (1959).

[30] M. S. Child, "Molecular Collision Theory," Academic Press, London, 1974.

[31] P. Pechukas, Phys. Rev. **181,** 166 (1969); **181,** 174 (1969).

[32] A. P. Penner and R. Wallace, Phys. Rev. **A7,** 1007 (1973); **11,** 149 (1975); G. Jolicard, J. Chem. Phys. **80,** 2476 (1984); G. D. Billing, J. Chem. Phys. **86,** 2617 (1987).

[33] F. Webster, P. J. Rossky and R. A. Friesner, Comp. Phys. Comm. **63,** 494 (1991); D. F. Coker and L. Xiao, J. Chem. Phys. **102,** 496 (1995).

[34] N. Makri and W. H. Miller, Chem. Phys. Lett. **151,** 1 (1981); N. Makri, Comp. Phys. Comm. **63,** 389 (1991).

[35] W. H. Miller, J. Chem. Phys. **53,** 1949, 3578 (1970); R. A. Marcus, Chem. Phys. Lett. **7,** 525 (1970).

[36] M. S. Child, "Semiclassical Mechanics with Molecular Applications," Clarendon Press, Oxford, 1991.

[37] J. H. van Vleck, Proc. Natl. Acad. Sci. U.S.A. **14,** 178 (1928).

[38] V. P. Maslov and M. V. Fedoriuk, "Semiclassical Approximation in Quantum Mechanics," Reidel, Boston, 1981.

[39] W. H. Miller, J. Chem. Phys. **95,** 9428 (1991).

[40] V. S. Filinov, Nucl. Phys. **B 271,** 717 (1986); N. Makri and W. H. Miller, Chem. Phys. Lett. **139,** 10 (1987); J. Chem. Phys. **89,** 2170 (1988); N. Makri, Comp. Phys. Comm. **63,** 389 (1991); B. W. Spath and W. H. Miller, J. Chem. Phys. **104,** 95 (1996).

[41] W. H. Press, B. P. Flannery, S. A. Teukolsky and W. T. Vetterling, "Numerical Recipes," Cambridge University Press, Cambridge, 1986.

[42] M. Abramovitz and I. A. Stegun, "Handbook of Mathematical Functions," Dover Publications, New York, 1965.

[43] Z. Bacic and J. C. Light, J. Chem. Phys. **85,** 4594 (1986); **87, 4008** (1987).

[44] R. Kosloff, Ann. Rev. Phys. Chem. **45,** 145 (1994).

[45] J. K. Cullum and R. A. Willoughby, "Lanczos Algorithms for Large Symmetric Eigenvalue Computations," Birkhaüser, Boston, 1985.

[46] D. Bohm, Phys. Rev. **85,** 166 (1952).

[47] F. H. M. Faisal and U. Schwengelbeck, Phys. Lett. A. **207,** 31 (1995).

[48] C. Eckart, Phys. Rev. **35,** 1303 (1930); H. S. Johnston and D. Rapp, J. Am. Chem. Soc. **83,** 1 (1961).

[49] D. R. Bates, "Quantum Theory, I. Elements," Academic Press, New York, 1961.

[50] D. Neuhauser and M. Baer, J. Chem. Phys. **90,** 4351 (1989).

[51] S. Brouard, D. Macias and J. G. Muga, J. Phys. A: Math. Gen. **27,** L439 (1994).

[52] G. Jolicard, Ann. Rev. Phys. Chem. **46,** 83 (1995).

[53] A. Kuppermann and G. C. Schatz, J. Chem. Phys. **62,** 2502 (1975); J. M. Launay and M. le Dourneuf, J. Phys. B. **15,** L455 (1982); D. E. Manolopoulos and R. E. Wyatt, Chem. Phys. Lett. **159,** 123 (1989); J. Z. H. Zhang and W. H. Miller, J. Chem. Phys. **91,** 1528 (1989); N. Blais, M. Zhao, D. G. Truhlar, D. W. Schwenke and D. J. Kouri, Chem. Phys. Lett. **166,** 166 (1990); S. L. Mielke, R. S. Friedman, D. G. Truhlar and D. W. Schwenke, Chem. Phys. Lett. **188,** 359 (1992); A. Kuppermann, J. Phys. Chem. **100,** 2621 (1996) and references therein.

[54] G. D. Billing, Chem. Phys. **146,** 423 (1990); **161,** 245 (1992).

[55] C. Eckart, Phys. Rev. **47,** 557 (1935).

[56] G. D. Billing, Chem. Phys. **161,** 245 (1992).

[57] W. H. Miller, N. C. Handy and J. E. Adams, J. Chem. Phys. **72, 99** (1980).

[58] G. D. Billing, Chem. Phys. **89,** 199 (1984).

[59] D. G. Truhlar, W. L. Hase and J. T. Hynes, J. Phys. Chem. **87,** 2664 (1983).

[60] E. P. Wigner, Z. Phys. Chem. **B19,** 203 (1932).

[61] E. Merzbacher, "Quantum Mechanics," John Wiley & Sons, Tokyo, 1961.

[62] R. A. Marcus and M. E. Coltrin, J. Chem. Phys. **67,** 2609 (1977).

[63] P. Pechukas and F. J. McLafferty, J. Chem. Phys. **58,** 1622 (1973).

[64] W. H. Miller, J. Chem. Phys. **76,** 4904 (1982).

[65] R. T. Skodje, D. G. Truhlar and B. C. Garrett, J. Chem. Phys. **77,** 5955 (1982); B. C. Garrett, D. G. Truhlar and G. C. Schatz, J. Am. Chem. Soc. **108,** 2876 (1986).

[66] W. H. Miller, J. Chem. Phys. **61,** 1823 (1974).

[67] W. H. Miller, S. D. Schwartz and J. W. Tromp, J. Chem. Phys. **79,** 4889 (1983).

[68] R. D. Levine, "Quantum Mechanics of Molecular Rate Processes," Oxford University Press, 1969.

[69] M. Shapiro and R. Bersohn, Ann. Rev. Phys. Chem. **33,** 409 (1982).

[70] D. P. Craig and T. Thirunamachandran, "Molecular Quantum Electrodynamics," Academic Press, New York, 1984.

[71] A. M. Arthurs and A. Dalgarno, Proc. R. Soc. London Ser. **A 256**, 540 (1960).

[72] D. M. Brink and G. R. Satchler, "Angular Momentum," Clarendon Press, Oxford, 1968.

[73] D. Brown and J. C. Light, J. Chem. Phys. **97**, 5465 (1992); W. H. Miller, J. Chem. Phys. **61**, 1823 (1974); W. H. Miller, S. D. Schwartz and J. W. Tromp, J. Chem. Phys. **79**, 4889 (1983); T. Seideman and W. H. Miller, J. Chem. Phys. **96**, 4412 (1992); T. Yamamoto, J. Chem. Phys. **33**, 281 (1960).

[74] W. H. Miller, J. Phys. Chem. **99**, 12387 (1995).

[75] E. Thiele, J. Chem. Phys. **45**, 491 (1966).

[76] J. C. Light, Disc. Faraday Soc. **44**, 14 (1967).

[77] R. D. Levine, "Quantum Mechanics of Molecular Rate Processes," Clarendon Press, Oxford, 1969.

[78] R. B. Bernstein and R. D. Levine, Adv. At. Mol. Phys. **11**, 215 (1975); R. D. Levine and J. L. Kinsey, in R. B. Bernstein, ed., "Atom Molecule Collison Theory: A Guide for the Experimentalist," Plenum, New York, 1979, p. 693.

[79] S. Augustin and H. Rabitz, J. Chem. Phys. **64**, 1223 (1976); **66**, 269 (1977); G. D. Billing, "Introduction to the Theory of Inelastic Collisions in Chemical Kinetics," Copenhagen, 1978.

[80] R. Loudon, "The Quantum Theory of Light," Clarendon Press, Oxford, 1983; R. Schinke, "Photodissociation Dynamics," Cambridge University Press, Cambridge, 1993.

[81] R. E. Wyatt and D. Scott, in J. Cullum and R. Willoughby, eds., "Large Eigenvalue Problems," North-Holland, Amsterdam, 1986; N. Moiseyev, R. A. Friesner and R. E. Wyatt, J. Chem. Phys. **85**, 331 (1986).

[82] A. Erdelyi, "Higher Transcendal functions," Vol. 2, McGraw-Hill, New York, 1953.

[83] I. S. Gradshteyn and I. M. Ryzhik, "Table of Integrals, Series, and Products," Academic Press, New York, 1965.

[84] R. C. Tolman, "The Principles of Statistical Mechanics," Oxford University Press, Oxford, 1938.

[85] D. A. McQuarrie, "Statistical Mechanics," Harper & Row, New York, 1976.

[86] R. Balian, "From Microphysics to Macrophysics," Springer-Verlag, Berlin, 1991.

[87] P. M. Morse, "Thermal Physics," W. A. Benjamin, New York, 1969.

[88] D. Chandler, "Introduction to Modern Statistical Mechanics," Oxford University Press, Oxford, 1987.

[89] N. G. Van Kampen, "Stochastic Processes in Physics and Chemistry," 2nd ed. Elsevier, Amsterdam, 1992.

[90] M. Born, Z. Phys. **1**, 45 (1920).

[91] J. G. Kirkwood, J. Chem. Phys. **2**, 351 (1934).

[92] J. G. Kirkwood and F. Westheimer, J. Chem. Phys. **6**, 506 (1936).

[93] L. Onsager, J. Am. Chem. Soc. **58**, 1486 (1936).

[94] E. McRae, J. Phys. Chem. **61**, 562 (1957).

[95] N. Bayliss and E. McRae, J. Phys. Chem. **58**, 1002 (1958).

[96]   C. J. F. Böttcher, "Theory of Electric Polarization," Elsevier, New York, 1952.

[97]   R. Pethig, "Dielectric and Electronic Properties of Biological Materials," John Wiley & Sons, New York, 1979.

[98]   J. Hylton, R. E. Christoffersen and G. G. Hall, Chem. Phys. Lett. **24**, 501 (1974).

[99]   D. L. Beveridge and G. W. Schnuell, J. Phys. Chem. **79**, 2562 (1975).

[100] M. D. Newton, J. Phys. Chem. **79**, 2795 (1975).

[101] J. O. Noell and K. Morokuma, Chem. Phys. Lett. **36**, 465 (1975).

[102] J. L. Rivail and D. Rinaldi, Chem. Phys. **18**, 233 (1976).

[103] A. Warshel, Chem. Phys. Lett. **55**, 454 (1978).

[104] D. Rinaldi, Comput. Chem. **6**, 155 (1982).

[105] E. S. Marcus, B. Terryn and J. L. Rivail, J. Phys. Chem. **87**, 4695 (1985).

[106] R. Contreras and A. Aizman, Int. J. Quant. Chem. **27**, 193 (1985).

[107] Y. Inone, H. Hoshi, M. Sakurai and R. Chujo, J. Chem. Phys. **87**, 1107 (1987).

[108] G. Karlström, J. Phys. Chem. **93**, 4952 (1989).

[109] O. Tapia, in H. Ratajczak and W. J. Orville-Thomas, eds., "Molecular Interactions," John Wiley & Sons, New York, 1980.

[110] O. Tapia, in R. Daudel, A. Pullman, L. Salem and A. Veillard, eds., "Quantum Theory of Chemical Reactions," Wiley, Dordrecht, 1980.

[111] K. V. Mikkelsen, E. Dalgaard and P. Swanstrøm, J. Phys. Chem. **79**, 587 (1987).

[112] K. V. Mikkelsen, H. Ågren, H. J. Aa. Jensen and T. Helgaker, J. Chem. Phys. **89**, 3086 (1988).

[113] K. V. Mikkelsen and M. Ratner, J. Chem. Phys. **90**, 4237 (1989).

[114] K. V. Mikkelsen and M. Ratner, J. Phys. Chem. **93**, 1759 (1989).

[115] C. Medina-Llanos, H. Ågren, K. Mikkelsen and H. J. Aa. Jensen, J. Chem. Phys. **90**, 6422 (1989).

[116] H. Ågren, C. Medina-Llanos, K. Mikkelsen and H. J. Aa. Jensen, Chem. Phys. Lett. **153**, 322 (1989).

[117] H. J. Kim and J. T. Hynes, J. Chem. Phys. **96**, 5088 (1992).

[118] R. Bianco, J. J. i Timoneda and J. T. Hynes, J. Phys. Chem. **98**, 12103 (1994).

[119] J. N. Gehlen, D. Chandler, H. J. Kim and J. T. Hynes, J. Phys. Chem. **96**, 1748 (1992).

[120] R. A. Marcus, J. Phys. Chem. **96**, 1753 (1992).

[121] R. H. Young, J. Chem. Phys. **97**, 8261 (1992).

[122] H. J. Kim, R. Bianco, B. J. Gertner and J. T. Hynnes, J. Phys. Chem. **97**, 1723 (1993).

[123] R. A. Marcus, J. Chem. Phys. **24**, 979 (1956).

[124] B. U. Felderhof, J. Chem. Phys. **67**, 493 (1977).

[125] S. Lee and J. T. Hynes, J. Chem. Phys. **88**, 6853 (1988).

[126] H. J. Kim and J. T. Hynes, J. Chem. Phys. **93**, 5194 (1990).

[127] J. Hylton, R. E. Christoffersen and G. G. Hall, Chem. Phys. Lett. **24**, 501 (1974).

[128] J. H. McCreery, R. E. Christoffersen and G. G. Hall, J. Am. Chem. Soc. **98**, 7191 (1976).

[129] O. Tapia and O. Goscinski, Mol. Phys. **29**, 1653 (1975).

[130] J-L. Rivail and D. Rinaldi, Chem. Phys. **18**, 233 (1976).

[131] S. Miertuus, E. Scrocco and J. Tomasi, Chem. Phys. **55**, 117 (1981).

[132] S. Miertuus and J. Tomasi, Chem. Phys. **65**, 239 (1982).

[133] K. V. Mikkelsen, E. Dalgaard and P. Swanstrøm, J. Phys. Chem. **91**, 3081 (1987).

[134] K. V. Mikkelsen, H. Ågren, H. J. Aa. Jensen and T. Helgaker, J. Chem. Phys. **89**, 3086 (1988).

[135] K. V. Mikkelsen, P. Jørgensen and H. J. Aa. Jensen, J. Chem. Phys. **100**, 6597 (1994).

[136] M. W. Wong, M. J. Frisch and K. B. Wilberg, J. Am. Chem. Soc. **113**, 4776 (1991).

[137] M. M. Karelson and M. C. Zerner, J. Phys. Chem. **96**, 6949 (1992).

[138] K. V. Mikkelsen, A. Cesar, H. Ågren and H. J. Aa. Jensen, J. Chem. Phys. **103**, 9010 (1995).

[139] J. Tomasi and M. Persico, Chem. Rev. **94**, 2027 (1994).

[140] M. W. Wong, K. B. Wiberg and M. J. Frisch, J. Am. Chem. Soc. **114**, 1645 (1992).

[141] K. B. Wiberg and M. W. Wong, J. Am. Chem. Soc. **115**, 1078 (1993).

[142] M. W. Wong and K. B. Wiberg, J. Chem. Phys. **95**, 8991 (1991).

[143] H. Ågren, S. Knuts, K. V. Mikkelsen and H. J. Aa. Jensen, Chem. Phys. **159**, 211 (1992).

[144] H. Hoshi, M. Sakurai, Y. Inoue and R. Chûjô, J. Chem. Phys. **87**, 1107 (1987).

[145] R. Cammi and J. Tomasi, J. Chem. Phys. **100**, 7495 (1994).

[146] M. Cossi, B. Mennucci and J. Tomasi, Chem. Phys. Lett. **228**, 165 (1994).

[147] H. Fröhlich, in "Theory of Dielectrics," Oxford, London, 1949.

[148] F. K. Fong, in "Theory of Molecular Relaxation," John Wiley & Sons, New York, 1975.

[149] R. Pething, in "Dielectric and Electronic Properties of Biological Materials," John Wiley & Sons, New York, 1979.

[150] M. A. Aguilar, F. J. Olivares del Valle and J. Tomasi, J. Chem. Phys. **98**, 7375 (1993).

[151] H. J. Aa. Jensen, H. Ågren and J. Olsen, in E. Clementi, ed., "MOTEC, Modern Techniques in Computational Chemistry," ESCOM, Leiden, 1991.

[152] H. Ågren, C. Medina-Llanos, K. V. Mikkelsen and H. J. Aa. Jensen, Chem. Phys. Lett. **153**, 322 (1988).

[153] C. Medina-Llanos, H. Ågren, K. M. Mikkelsen and H. J. Aa. Jensen, J. Chem. Phys. **90**, 6422 (1989).

[154] G. Herzberg, "Infrared and Raman Spectra," Van Nostrand Reinhold, New York, 1956.

[155] A. Cesar, H. Ågren, T. Helgaker, P. Jørgensen and H. J. Aa. Jensen, J. Chem. Phys. **45**, 5906 (1991).

[156] K. V. Mikkelsen, Z. Phys. Chem. **170**, 129 (1991).

[157] K. V. Mikkelsen and M. Ratner, Int. J. Quant. Chem. **21**, 341 (1987).

[158] K. V. Mikkelsen, P. Jørgensen and H. J. Aa. Jensen, J. Chem. Phys. **100**, 6597 (1994).

[159] K. V. Mikkelsen, Y. Luo, H. Ågren and P. Jørgensen, J. Chem. Phys. **100**, 8240 (1994).

[160] J. Olsen and P. Jørgensen, J. Chem. Phys. **82**, 3235 (1985).

[161] K. D. Singer and A. F. Garito, J. Chem. Phys. **75**, 3572 (1981).

[162] P. N. Prasad and D. J. Williams, "Introduction to Nonlinear Optical Effects in Organic Molecules and Polymers," John Wiley & Sons, New York, 1991.

[163] R. W. Boyd, "Nonlinear Optics," Academic Press, New York, 1992.

[164] H. A. Lorentz, "The Theory of Electrons," Dover, New York, 1952.

[165] D. M. Bishop, Rev. Mod. Phys. **62**, 343 (1990).

[166] In D. N. Zubarev, ed., Nonlinear Statistical Thermodynamics, Plenum, New York, 1974.

[167] D. J. Thouless, "The Quantum Mechanics of Many-Body Systems," Academic Press, New York, 1961.

[168] E. Dalgaard, Phys. Rev. A **26**, 42 (1982).

[169] D. L. Yeager and P. Jørgensen, Chem. Phys. Lett. **65**, 77 (1979).

[170] E. Dalgaard, J. Chem. Phys. **72**, 816 (1980).

[171] C. Funchs, V. Bonacic-Koutecky and J. Koutecky, J. Chem. Phys. **98**, 3121 (1993).

[172] S. Kotoku, F. Aiga and R. Itoh, J. Chem. Phys. **99**, 3738 (1993).

[173] E. S. Nielsen, P. Jørgensen and J. Oddershede, J. Chem. Phys. **73**, 6238 (1980).

[174] P. E. S. Wormer and W. Rijks, Phys. Rev. A **75**, 2928 (1986).

[175] J. E. Rice and N. Handy, J. Chem. Phys. **94**, 4959 (1991).

[176] J. E. Rice and N. Handy, Int. J. Quant. Chem. **43**, 91 (1992).

[177] F. Aiga, K. Sasagane and R. Itoh, J. Chem. Phys. **99**, 3779 (1993).

[178] H. J. Monkhorst, Int. J. Quant. Chem. **11**, 421 (1977).

[179] E. Dalgaard and H. J. Monkhorst, Phys. Rev. A **28**, 1217 (1983).

[180] H. Koch and P. Jørgensen, J. Chem. Phys. **93**, 3333 (1990).

[181] K. V. Mikkelsen and K. O. Sylvester-Hvid, J. Phys. Chem. **100**, 11429 (1996).

[182] P. Jørgensen, H. J. Aa. Jensen and J. Olsen, J. Chem. Phys. **89**, 3654 (1988).

[183] J. Olsen, H. J. Aa. Jensen and P. Jørgensen, J. Comp. Chem. **74**, 265 (1988).

[184] K. V. Mikkelsen, P. Jørgensen and H. J. Aa. Jensen, J. Chem. Phys. **100**, 6597 (1994).

[185] K. V. Mikkelsen, Y. Luo, H. Ågren and P. Jørgensen, J. Chem. Phys. **1020**, 9362 (1995).

[186] T. Helgaker and P. Jørgensen, J. Chem. Phys. **95**, 2595 (1991).

[187] K. Ruud, T. Helgaker, K. L. Bak, P. Jørgensen and H. J. Aa. Jensen, J. Chem. Phys. **99**, 3847 (1993).

[188] K. Ruud, T. Helgaker, R. Kobayashi, P. Jørgensen, K. L. Bak and H. J. Aa. Jensen, J. Chem. Phys. **100**, 8178 (1994).

[189] K. Ruud, T. Helgaker, K. L. Bak, P. Jørgensen and J. Olsen, Chem. Phys. **195**, 157 (1995).

[190] F. London, J. Phys. Radium **8**, 397 (1937).

[191] K. V. Mikkelsen, K. Ruud, T. Helgaker and P. Jørgensen, J. Chem. Phys. (in press).

[192] K. V. Mikkelsen, K. Ruud and T. Helgaker, Chem. Phys. Lett. **253**, 443 (1996).

[193] G. C. Schatz and M. A. Ratner, "Quantum Mechanics in Chemistry," Prentice Hall, Englewood Cliffs, 1993.

[194] J. Ulstrup, "Charge Transfer Processes in Condensed Media," Sprigner-Verlag, Berlin-Heidelberg, 1979.

[195] J. Ulstrup and J. Jortner, J. Chem. Phys. **63**, 4358 (1975).

[196] N. R. Kestner, J. Logan and J. Jortner, J. Chem. Phys. **78**, 2148 (1974).

[197] R. P. Van Duyne and S. F. Fischer, Chem. Phys. **5**, 183 (1974).

[198] R. P. Van Duyne and S. F. Fischer, Chem. Phys. **26**, 9 (1977).

[199] P. P. Schmidt, J. Chem. Soc. Faraday II **69**, 1104 (1973).

[200] C. B. Duke and R. J. Meyer, J. Phys. Rev. B: Condens, Matter **B23**, 2118 (1981).

[201] J. G. Kirkwood, J. Chem. Phys. **2**, 351 (1934).

[202] J. G. Kirkwood and F. H. Westheimer, J. Chem. Phys. **6**, 506 (1938).

[203] R. A. Marcus, J. Chem. Phys. **24**, 966 (1956).

[204] R. A. Marcus, J. Chem. Phys. **24**, 979 (1956).

[205] R. A. Marcus, J. Chem. Phys. **43**, 679 (1965).

[206] V. Levich, Adv. Electrochem. Electrochem. Eng. **4**, 249 (1966).

[207] R. R. Doganadze and A. A. Kornishev, Phys. Status Solidi B **54**, 439 (1972).

[208] R. R. Doganadze and A. A. Kornishev, J. Chem. Soc. Faraday II **70**, 1121, (1974).

[209] V. G. Levich and R. R. Doganadze, Collect Czech. Chem. Commun. **26**, 193 (1961).

[210] Faraday Discuss. Chem. Soc. 74 (1982).

[211] M. D. Newton, ACS Symp. Ser. **189**, 254 (1982).

[212] M. D. Newton, Int. J. Quantum Chem. Symp. **14**, 363 (1980).

[213] N. S. Hush, in D. B. Brown, ed., "Mixed-Valence Compounds," Reidel, Dordrecht, 1980, p. 151.

[214] P. O. Löwdin, J. Chem. Phys. **18**, 365 (1950).

[215] M. D. Newton and N. Sutin, Ann. Rev. Phys. Chem. **35**, 437 (1984).

[216] R. R. Doganadze and A. M. Kuznetsov, Elektrokhimiya **2**, 1324 (1967).

[217] R. R. Doganadze, A. M. Kuznetsov and V. G. Levich, Electrochim. Acta **13**, 1025 (1968).

[218] M. A. Vorotyntsev, R. R. Doganadze and A. M. Kuznetsov, Dokl. Akad. Nauk SSSR **195**, 1135 (1970).

[219] I. Webman and N. R. Kestner, J. Phys. Chem. **83**, 451 (1979).

[220] J. D. Jackson, "Classical Electrodynamics," John Wiley & Sons, New York, 1964.

[221] E. Dalgaard, J. Phys. B: At. Mol. Phys. **9**, 2573 (1976).

[222] P. Ehrenfest, Z. Phys. **45**, 455 (1927).

[223] M. D. Newton and N. Sutin, Ann. Rev. Phys. Chem. **35**, 437 (1984).

[224] N. Sutin, Prog. Inorg. Chem. **30,** 441 (1983).

[225] R. A. Marcus and N. Sutin, Biochim. Biophys. Acta **811,** 265 (1985).

[226] N. S. Hush, Coord. Chem. Rev. **64,** 135 (1985).

[227] J. Phys. Chem. **90,** 3657 (1986).

[228] K. V. Mikkelsen and M. A. Ratner, Chem. Rev. **87,** 113 (1987).

[229] R. R. Dogonadze, A. M. Kuznetsov and T. A. Maragishvili, Electrochim. Acta **25,** 1 (1980).

[230] H. M. McConnell, J. Chem. Phys. **35,** 508 (1961).

[231] E. Heilbronner and A. Schmelzer, Helvetica Chim. Acta **58,** 936 (1975).

[232] M. N. Paddon-Row, S. S. Wong, and K. D. Jordan, J. Am. Chem. Soc. **112,** 1710 (1990).

[233] M. N. Paddon-Row, S. S. Wong and K. D. Jordan. J. Chem. Soc., Perkin Trans. II, 417 (1990).

[234] S. Larsson, J. Phys. Chem. **88,** 1321 (1984).

[235] A. Broo and S. Larsson, Int. J. Quant. Chem. Quant. Biol. Symp. **16,** 185 (1989).

[236] C. Joachim, J. P. Launay and S. Woitellier, Chem. Phys. **147,** 131 (1990).

[237] C. Joachim, J. P. Launay and S. Woitellier, Chem. Phys. **131,** 481 (1989).

[238] J. R. Reimers and N. S. Hush, Chem. Phys. **146,** 105 (1990).

[239] M. D. Newton, Chem. Rev. **91,** 767 (1991).

[240] K. D. Jordan and M. N. Paddon-Row, Chem. Rev. **92,** 395 (1992).

[241] K. D. Jordan and M. N. Paddon-Row, J. Phys. Chem. **96,** 1188 (1992).

[242] L. A. Curtiss, C. A. Naleway and J. R. Miller, Chem. Phys. **176,** 387 (1993).

[243] C. Liang and M. D. Newton, J. Phys. Chem. **97,** 199 (1993).

[244] R. Kosloff and M. A. Ratner, Israel J. Chem. **30,** 45 (1990).

[245] M. A. Ratner, J. Phys. Chem. **94,** 4877 (1990).

[246] J. N. Onuchic and D. N. Beratan, J. Chem. Phys. **92,** 722 (1990).

[247] A. Farazdel, M. Dupuis, E. Clementi and A. Aviram, J. Am. Chem. Soc. **112,** 4206 (1990).

[248] A. Broo and S. Larsson, Chem. Phys. **148,** 103 (1990).

[249] K. Ohta, O. Okada and T. Yoshida, J. Phys. Chem. **93,** 932 (1989).

[250] M. D. Newton, J. Phys. Chem. **92,** 3049 (1988).

[251] K. Ohta and K. Morokuma, J. Phys. Chem. **91,** 401 (1987).

[252] R. J. Cave, D. V. Baxter, W. A. Goodard III and J. D. Baldeschwieler, J. Chem. Phys. **87,** 926 (1987).

[253] K. Ohta, G. L. Closs, K. Morokuma and N. J. Green, J. Am. Chem. Soc. **108,** 1319 (1986).

[254] M. D. Newton, J. Phys. Chem. **90,** 3734 (1986).

[255] J. Logan and M. D. Newton, J. Chem. Phys. **78,** 4086 (1983).

[256] M. D. Newton, Int. J. Quant. Chem. Symp. **14,** 363 (1980).

[257] S. Larsson, A. Broo, B. Källebring and A. Volosov, Int. J. Quant. Chem. Biol. Symp. **15,** 1 (1988).

[258] M. D. Todd, K. V. Mikkelsen, J. T. Hupp and M. A. Ratner, New J. Chem. **15,** 97 (1991).

[259] K. V. Mikkelsen. Dissertation, Electron Transfer in Solution, Aarhus University, Århus, Denmark, 1985.

[260] K. V. Mikkelsen, E. Dalgaard and P. Swanstrøm, J. Phys. Chem. **91,** 3081 (1987).

[261] K. V. Mikkelsen and M. A. Ratner, Int. J. Quantum. Chem. Symp. **21,** 341 (1987).

[262] K. V. Mikkelsen and M. A. Ratner, Int. J. Quantum. Chem. Symp. **22,** 707 (1988).

[263] K. V. Mikkelsen and M. A. Ratner, J. Phys. Chem. **93,** 1759 (1989).

[264] K. V. Mikkelsen and M. A. Ratner, J. Chem. Phys. **90,** 4237 (1989).

[265] K. V. Mikkelsen, J. Phys. Chem., submitted.

[266] W. J. Pietro, M. M. Francl, W. J. Hehre, D. J. DeFrees, J. A. Pople and J. S. Binkley, J. Am. Chem. Soc. **104,** 5039 (1982).

[267] M. S. Gordon, J. S. Binkley, J. A. Pople, W. J. Pietro and W. J. Hehre, J. Am. Chem. Soc. **104,** 2797 (1982).

[268] J. S. Binkley, J. A. Pople and W. J. Hehre, J. Am. Chem. Soc. **102,** 939 (1980).

[269] P. C. Hariharan and J. A. Pople, Mol. Phys. **27,** 209 (1974).

[270] W. J. Hehre, R. Ditchfield and J. A. Pople, J. Chem. Phys. **56,** 2257 (1972).

[271] R. Ditchfield, W. J. Hehre and J. A. Pople, J. Chem. Phys. **54,** 724 (1971).

[272] S. Huzinaga, J. Chem. Phys. **42,** 1293 (1965).

[273] T. H. Dunning and P. J. Hay, in H. F. Schaeffer, ed., "Methods of Electronic Structure Theory," Plenum Press, New York, 1977, p. 1.

[274] J. Paldus and J. Cízek, J. Chem. Phys. **52,** 2919 (1970).

[275] J. Paldus and J. Cízek, Chem. Phys. Lett. **3,** 1 (1969).

[276] P. O. Löwdin, Phys. Rev. **97,** 1474 (1955).

[277] P. O. Löwdin, Phys. Rev. **97,** 1490 (1955).

[278] P. O. Löwdin, Phys. Rev. **97,** 1509 (1955).

[279] T. Koopmans, Physica (Utrecht) **1,** 104 (1933).

[280] J. N. Onuchic, D. N. Beratan, J. R. Winkler and H. B. Gray, Annu. Rev. Biophys. Biomol. Struc. **21,** 349 (1992).

[281] H. M. McConnell, J. Chem. Phys. **35,** 32 (1961).

[282] P. W. Anderson, Phys. Rev. **79,** 350 (1950).

[283] J. Halpern and L. E. Orgen, Discuss. Faraday Soc. **29,** 32 (1960).

[284] S. Larsson, J. Phys. Chem. **88,** 1321 (1984).

[285] S. Larsson, Chem. Scrip. **28,** 15 (1988).

[286] K. V. Mikkelsen, L. K. Skov, H. Nar and O. Farver, Proc. Natl. Acad. Sci. U.S.A. **90,** 5443 (1993).

[287] H. E. M. Christensen, L. S. Conrad, K. V. Mikkelsen, M. K. Nielsen and J. Ulstrup, Inorg. Chem. **29,** 2808 (1990).

[288] M. A. Ratner, J. Phys. Chem. **94,** 4977 (1990).

[289] A. S. S. da Gama, Quimica Nova **11,** 76 (1988).

[290] H. E. M. Christensen, L. S. Conrad, K. V. Mikkelsen and J. Ulstrup, J. Phys. Chem. **96,** 4451 (1992).

[291] P. Meystre and M. Sargent III, "Elements of Quantum Optics," 2nd ed., Springer-Verlag, Berlin, 1990.

[292] C. Cohen-Tannoudji, J. Dupont-Roc and G. Grynberg, "Atom–Photon Interactions," Wiley-Interscience, New York, 1992.

[293] S. Haroche, in J. Dalibard, J.-M. Raimond and J. Zinn-Justin, eds., "Fundamental Systems in Quantum Optics," North-Holland, The Netherlands, 1992.

[294] C. Cohen-Tannoudji, in J. Dalibard, J.-M. Raimond and J. Zinn-Justin, eds., "Fundamental Systems in Quantum Optics," North-Holland, The Netherlands, 1992.

[295] C. Cohen-Tannoudji, in G. Grynberg and R. Stora, eds., "New Trends in Atomic Physics," North-Holland, The Netherlands, 1984.

[296] J. P. Connerade, K. Conen, K. Dietz and J. Henkel, J. Phys. B: At. Mol. Phys. **25,** 3771, 1992.

[297] K. Codling and L. J. Frasinski, J. Phys. B: At. Mol. Phys. **26,** 783 (1993).

[298] D. P. Hodgkinson and J. S. Briggs, J. Phys. B: At. Mol. Phys. **10,** 2583 (1977).

[299] V. V. Korobkin, A. V. Borovskii and Ch. K. Mukhtarov, Laser Physics **3,** 999 (1993).

[300] K. Codling, L. J. Frasinski and P. A. Hatherly, J. Phys. B: At. Mol. Phys. **22,** 321 (1989).

[301] M. V. Fedorov, Physics Letters **61A,** 224 (1977).

[302] M. Mohan and P. Chand, Physics Letters **68A,** 45 (1978).

[303] U. Devi and M. Mohan, Physics Letters **53A,** 421 (1975).

[304] N. Saito and G. G. Hall, Int. J. Quant. Chem. **35,** 283 (1989).

[305] S. C. Leasure, K. F. Milfeld and R. E. Wyatt, J. Chem. Phys. **74,** 6197 (1981).

[306] T. Tung Nguyen-Dang, J. Chem. Phys. **90,** 2657 (1989).

[307] S.-I. Chu, J. V. Tietz and K. K. Datta, J. Chem. Phys. **77,** 2968 (1982).

[308] S. Chelkowski and A. D. Bandrauk, Chem. Phys. Lett. **186,** 264 (1991).

[309] S. L. Chin, Y. Liang, J. E. Decker, F. A. Ilkov and M. V. Ammosov, J. Phys. B: At. Mol. Phys. **25,** 249 (1992).

[310] V. N. Bagratashvili, V. S. Letokhov, A. A. Makarov and E. A. Ryabov, "Multiple Photon Infrared Laser Photophysics and Photochemistry," Harwood Academic Publishers, Chur, 1985.

[311] B. Sharma and M. Mohan, J. Phys. B: At. Mol. Phys. **25,** 399 (1992).

[312] P. L. Knight. "Notes from Erasmus Advanced Summer School on Quantum Optics," Crete, 1993.

[313] J. Linderberg and Y. Öhrn. "Propagators in Quantum Chemistry," Academic Press, London and New York, 1973.

[314] J. J. Sakurai, "Advanced Quantum Mechanics," Addison-Wesley, Redwood City, CA, 1967.

[315] J. Providencia, Nucl. Phys. B **57,** 536 (1973).

[316] K. V. Mikkelsen and M. Kmit, Theor. Chim. Acta **90,** 307 (1995).

ort2

2rt

[317] G. D. Billing and K. V. Mikkelsen, "Introduction to Molecular Dynamics and Chemical Kinetics," John Wiley & Sons, New York, 1996.

[318] R. P. Bell, "The Proton in Chemistry," Chapman and Hall, London, 1973.

[319] R. A. Marcus, J. Chem. Phys. **72,** 892 (1968).

[320] D. Borgis, S. Lee and J. T. Hynes, Chem. Phys. Lett. **162,** 19 (1989).

[321] D. Borgis and J. T. Hynes, J. Chem. Phys. **94,** 3619 (1991).

[322] D. Borgis and J. T. Hynes, Chem. Phys. **170,** 315 (1993).

[323] T. Yamamoto, J. Chem. Phys. **33,** 281 (1960).

[324] D. Chandler, J. Chem. Phys. **68,** 2959 (1978).

[325] S. H. Northrup and J. T. Hynes, J. Chem. Phys. **73,** 2700 (1980).

[326] G. Voth, D. Chandler and W. H. Miller, J. Chem. Phys. **91,** 7749 (1989).

INDEX

RETURN TO ➡ PHYSICS LIBRARY
351 LeConte Hall
642-3122

LOAN PERIOD 1	2	3
1-MONTH		
4	5	6

ALL BOOKS MAY BE RECALLED AFTER 7 DAYS
Overdue books are subject to replacement bills

DUE AS STAMPED BELOW

This book will be held in PHYSICS LIBRARY until OCT 12 1998		
NOV 2 0 1998		
DEC 1 5 2000		
OCT 3 0 2006		
MAY 1 0 2010		
JUL 0 5 2011		

UNIVERSITY OF CALIFORNIA, BERKELEY
BERKELEY, CA 94720

FORM NO. DD 25